SpringerBriefs in Molecular Science

Biometals

Series Editor
Larry L. Barton

For further volumes:
http://www.springer.com/series/10046

Dominique Expert · Mark R. O'Brian
Editors

Molecular Aspects of Iron Metabolism in Pathogenic and Symbiotic Plant–Microbe Associations

 Springer

Editors
Dominique Expert
Centre National de la Recherche
  Scientifique
Laboratoire Interactions
  Plantes-Pathogènes, AgroParisTech
Paris
France

Mark R. O'Brian
Department of Biochemistry
State University of New York at Buffalo
Buffalo, NY
USA

ISSN 2212-9901
ISBN 978-94-007-5266-5   ISBN 978-94-007-5267-2   (eBook)
DOI 10.1007/978-94-007-5267-2
Springer Dordrecht Heidelberg New York London

Library of Congress Control Number: 2012944980

Printed on acid-free paper

Springer is part of Springer Science+Business Media (www.springer.com)

Contents

Contributors

Alia Dellagi Département Sciences de la Vie et Santé, Laboratoire Interactions Plantes-Pathogènes, AgroParisTech, 16 rue Claude Bernard, 75005 Paris, France

Dominique Expert Centre National de la Recherche Scientifique, Laboratoire Interactions Plantes-Pathogènes, AgroParisTech, Paris, France

Elena Fabiano Departmento de Bioquímica y Genómica Microbianas, Instituto de Investigaciones Biológicas Clemente Estable, Av. Italia 3318, 11600 Montevideo, Uruguay

Thierry Franza Institut National de la Recherche Agronomique, Laboratoire Interactions Plantes-Pathogènes, AgroParisTech, 16 rue Claude Bernard, 75005 Paris, France

Mark R. O'Brian Department of Biochemistry, State University of New York at Buffalo, 140 Farber Hall, Buffalo, NY 14214, USA

Chapter 1
Iron, an Element Essential to Life

Dominique Expert

Abstract Iron plays a vital role in virtually all living organisms. This element is the second most common metal after aluminum in the earth's crust. Its abundance and the flexibility of its electronic structure made iron particularly suitable for life. Indeed, the Fe^{3+}/Fe^{2+} couple covers a wide range of redox potentials which can be finely tuned by coordinated ligands, conferring on it a key catalytic role in various fundamental metabolic pathways. However, as ferrous iron catalyzes the production of cell-damaging reactive oxygen speices $OH°$ via the Fenton reaction, excess iron or incorrect storage of this metal can be deleterious to organisms. Despite its abundance, iron is not easily bioavailable under aerobic conditions because the oxidized ferric form displays low solubility. Confronted with shortages of iron, organisms with aerobic lifestyles express specific mechanisms for its acquisition. Thus, iron is often a stake in competition between organisms of the same ecological niche and holds a peculiar position at the microbe–host interface. This chapter illustrates the importance of this metal in biological systems.

Keywords Iron biochemistry history · Iron biology history

Iron is a trace element essential to almost every living cell, in microbes, plants, and animals. Vital processes such as photosynthesis, respiration, DNA synthesis, nitrogen fixation, and detoxification of free radicals depend on the activity of iron-containing enzymes and proteins. Proteins using iron as a metal cofactor display great diversity. Enzymes containing iron play important roles in the basic biochemical mechanisms, proteins containing iron–sulfur clusters, or heme mediate redox and electron transfer reactions. Hemoglobin and leghemoglobin present in

D. Expert (✉)
Centre National de la Recherche Scientifique, Laboratoire Interactions
Plantes-Pathogènes, AgroParisTech, 16 rue Claude Bernard, 75005 Paris, France
e-mail: expert@agroparistech.fr

D. Expert and M. R. O'Brian (eds.), *Molecular Aspects of Iron Metabolism in Pathogenic and Symbiotic Plant–Microbe Associations*, SpringerBriefs in Biometals,
DOI: 10.1007/978-94-007-5267-2_1, © The Author(s) 2012

1

root nodules of nitrogen-fixing plants provide another representative class of iron proteins which bind oxygen. Why is iron involved in so many metabolic processes? The catalytic function of iron relies on its electronic structure. Its position in the middle of the elements in the first transition series implies that it can exist in various oxidation states, principally ferrous (Fe^{2+}) and ferric (Fe^{3+}). In addition, it can undergo reversible changes in its oxidation states which differ by one electron. Remarkably, its redox properties can be modified by its ligand environment and fine-tuning by well-adapted coordinated ligands makes iron-containing enzymes to display redox potentials able to cover a wide range of nearly 1 V. Owing to the flexibility of its electronic equipment, the iron atom is thought to have played a pivotal role in the history of the earliest Earth ecosystems. In this introduction, we wish to illustrate the importance of iron in biological systems by briefly depicting the evolutional path of this element through the geological times.

Iron is the fourth most abundant element in the Earth's crust after oxygen, silicium, and aluminum, making up 4.7 % of the total crust mass (Williams and Fraùsto da Silva 1999). Because of its abundance in the prebiotic world, iron is believed to have been selected as catalyst in the former energy producing chemical reactions. Interestingly, for the authors of the theory of a chemoautotrophic origin of life, who consider that the prime source of energy for carbon fixation is of chemical nature, the necessary reducing power arises from the oxidative formation of pyrite (FeS_2) from ferrous sulfide (FeS) and hydrogen sulfide (H_2S) (Wachtershaüser 1990; Huber and Wachtershaüser 1997). In this world, that is 4–2.4 billion years ago, the atmosphere was anoxic and essentially composed of nitrogen, carbon dioxide, and water (Williams and Fraùsto da Silva 1999). Iron was present in the reduced state, quite soluble, and thus available for life. But the idea that oxidation of ferrous iron was taking place naturally, in the absence of free oxygen, was a matter of controversy (Canfield et al. 2006). The discovery of ferric oxides contained in banded iron formations (BIF) dated from the Archean and Proterozoic ages (3.5–2.5 billions years ago) was in this respect of prime interest (Dietrich et al. 2006). It thus became plausible that anaerobic ferrous iron oxidation could have occurred before the evolution of oxygenic photosynthesis. Such a hypothesis was strengthened by the discovery of purple, nonsulfur, phototrophic bacteria able to oxidize Fe^{2+} to Fe^{3+} and reduce carbon dioxide to organic matter in the absence of oxygen (Widdel et al. 1993). Finding nitrate-reducing bacteria which gain energy for growth by oxidizing ferrous iron in an anaerobic manner was also considered a convincing argument (Straub et al. 2001).

With the rise of oxygen into the biosphere, around 2.3 billions years ago, there was a considerable change in the redox balance on Earth (Williams and Fraùsto da Silva 1999). It is believed that oxygenic photosynthesis evolved before the atmosphere became permanently oxygenated. One consequence was a progressive transformation in the availability of elements with certainly dramatic effects on anaerobic indigenous populations (Raymond and Segrè 2006). Very likely, sulfide and ferrous iron were the first reducing chemicals to have been removed from the ocean by dioxygen: they were oxidized to sulfate and ferric ion, the latter precipitating as ferric hydroxides. In this regard, the stalk-like morphologies observed in contemporary lithotrophic iron-oxidizing bacteria, also identified in filamentous Fe microfossils might be important markers of the Earth's oxygen history (Chan et al. 2011). Indeed, these bacteria are present in freshwater and marine environments in which redox gradients of oxygen and ferrous iron exist. Analysis of the stalk of *Mariprofundus ferroxydans* revealed the formation of a biomineralized structure containing carboxyl-rich polysaccharides and granular iron oxyhydroxides (Singer et al. 2011). Excreted from the cell as fibrils, these structures are supposed to enhance the elimination of the ferric waste product of Fe-oxidation metabolism.

Ferric hydroxides are insoluble at pH > 4. The solubility constant Ksp of $Fe(OH)_3$ is 10^{-38}. At pH 7, Fe^{3+} is available at 10^{-17}M, which is far below the micromolar concentrations required for microbial growth (Neilands 1991). The solubility of iron decreases three orders of magnitude per pH unit. In soils, iron is present as insoluble hydrated ferric oxides commonly described as rust (Loeppert et al. 1994). Their dissolution takes place by reduction or complexation, organic components present in the rhizosphere playing a major role in these processes (Lindsay and Schwab 1982). However, about 30 % of croplands are too alkaline for optimal plant growth. Confronted with a lack of iron availability, living organisms have developed adapted mechanisms to acquire this metal from their environment. They elaborated high-affinity uptake systems, based on the expression of plasma membrane-bound ferric reductases or the production of ferric-specific chelating molecules. For instance, fungi and plants use ferric reduction (reviewed in Kornitzer 2009; Labbé et al. 2007; Morrissey and Guerinot 2009). In body fluids of vertebrates and some invertebrates, mainly worms and insects, ferric iron is bound to transferrin or transferrin-like proteins (reviewed in Gkouvatsos et al. 2011). Microorganisms and certain plants such as grasses excrete sidero-phores, which are high affinity Fe^{3+} scavenging/solubilizing small molecules that, once loaded with iron, are specifically imported into cells (Neilands 1995; Kobayashi et al. 2010; Krewulak and Vogel 2008, see Fig.1). Microbial sidero-phores vary widely in their overall structure, which accounts for the specific recognition and uptake by a given microorganism, but the iron-chelating functional groups, catechol, hydroxamate, and carboxylate, are well conserved (Budzikiewicz 2010).

In aerobic environments, iron catalyzes the single electron reduction of oxygen giving rise to oxidizing radicals, which may be very damaging to biomolecules. Iron toxicity is involved in lipid peroxidation, protein degradation, and DNA mutations. Living cells protect themselves by strictly controlling their intracellular concentration of iron, which requires the coordinated regulation of the synthesis and action of proteins involved in acquisition, utilization, and storage of the metal. When present in excess, iron is stored in a nontoxic form, in ferritins. Ferritins constitute a remarkable family of iron proteins, widespread in all domains of life. They are organized in a 24-subunit shell surrounding a central cavity, and owing to their ferroxidase activity, they oxidize excess of ferrous ions and store the ferric form in a bioavailable mineral core (Theil 2003; Briat et al. 2010; Andrews 1998). Ferritin is considered as an ancient protein which has evolved to solve the problem of iron/oxygen chemistry and metabolism. When intracellular levels are low, the cell is able to increase its acquisition of iron by either mobilizing stored iron or utilizing external sources. The cell has also the ability to prioritize its iron utilization so that iron-containing proteins preferentially receive iron, at the same time avoiding toxic side reactions. For instance, biosynthesis of iron–sulfur proteins involves complex protein machineries that build iron–sulfur clusters from intracellular iron and sulfur and transfer them to their cognate protein acceptors (Xu and Møller 2011). Present in all organisms, these machineries protect the cellular surroundings from the potentially deleterious effects from free iron and sulfur. Numerous studies focused on diverse organisms, from bacteria to humans, point to the existence of proteins acting both as iron sensors and as regulators of gene expression for functions participating in the maintenance of iron homeostasis (Hentze et al. 2010; Kaplan and Kaplan 2009; Ivanov et al. 2012). The best studied iron-responsive transcriptional regulator in Gram-negative bacteria certainly is the ferric uptake regulator (Fur) protein which controls the expression of iron acquisition and storage systems (Hantke 2001). Fur protein acts as a dimer, each monomer containing a non-heme ferrous iron site. If the cellular iron level becomes too low, the active Fur repressor loses Fe^{2+}, its co-repressor, and is no longer able to bind to its operator sites. Interestingly, it was found that genes other than those involved in iron homeostasis can be Fur-regulated. Depending on the bacterial niche considered, Fur can regulate genes involved in pathogenicity, redox-stress resistance, or energy metabolism. However, the Fur family of bacterial iron regulatory proteins is not unique. In rhizobia for instance, new iron sensory proteins were identified that do not strictly conform to the Fur paradigm (Rudolph et al. 2006).

Mainly present as a chelated element, iron often constitutes a growth-limiting factor and the stake in an ardent competition between various members of a particular ecological niche or habitat. In the rhizosphere, siderophores released by bacteria and fungi can capture iron from natural chelates, thus depriving of iron microorganisms that produce siderophores in lower concentrations or with a lower affinity for this metal (Kloepper et al. 1980; Lemanceau and Alabouvette 1993; Haas and Defago 2005; Robin et al. 2007). Another example of iron competition resides in pathogenic microorganisms that invade vertebrate hosts. The topic of iron and infection arose when Schade and Caroline (1944), discovered the presence of transferrin in blood and ovotransferrin in egg whites, and noted that these proteins inhibited the growth of certain bacteria. Close attention has then been drawn to iron as a clue to bacterial virulence and animal host defense (Bullen 1981; Payne 1989; Litwin and Calderwood 1993; Schaible and Kaufmann 2004; Weinberg 2009). Host antimicrobial mechanisms include the well-known iron-withholding strategy and are considered as a part of the vertebrate innate immune system (Ong et al. 2006; Nairz et al. 2010). The problem of iron availability and toxicity for plant pathogens and rhizobial symbionts has been raised more recently (Expert and Gill 1991; Leong and Neilands 1982; Expert et al. 1996; Franza and Expert 2010; O'Brian and Fabiano 2010). Knowledge of strategies of iron acquisition and control of iron homeostasis in microorganisms living in close associations with plants has progressed considerably in recent years (Barton and Abadia 2006). Much attention has also been turned to molecular mechanisms involved by plants to cope with iron deficiency. These studies have opened perspectives to further understanding the effects of iron in plant–microbe interactions. In the following chapters, we will consider the role of iron in microbial virulence, plant defense, and symbiotic legume associations with particular attention to *Erwinia*, *Rhizobium*, and related genera, the studies of which have most contributed to broaden the concept of iron modulators in plant–microbe interactions.

References

Andrews SC (1998) Iron storage in bacteria. Adv Microb Physiol 40:281–351

Barton LB, Abadia J (2006) Iron nutrition in plants and rhizospheric microorganisms. Springer, Dordrecht

Briat JF, Duc C, Ravet K, Gaymard F (2010) Ferritins and iron storage in plants. Biochim Biophys Acta 1800:806–814

Bullen JJ (1981) The significance of iron in infection. Rev Infect Dis 3:1127–1138

Budzikiewicz H (2010) Microbial siderophores. Fortschr Chem Org Naturst 92:1–75

Canfield DE, Rosing MT, Bjerrum C (2006) Early anaerobic metabolisms. Phil Trans R Soc B 361:1819–1836

Chan CS, Fakra SC, Emerson D, Fleming EJ, Edwards KJ (2011) Lithotrophic iron-oxidizing bacteria produce organic stalks to control mineral growth: implications for biosignature formation. ISME J 5:717–727

Dietrich LE, Tice MM, Newman DK (2006) The co-evolution of life and earth. Curr Biol 16:R395–R400

Expert D, Enard C, Masclaux D (1996) The role of iron in pathogenic plant-microbe interactions. Trends Microbiol 4:232–236

Expert D, Gill PR (1991) Iron: a modulator of bacterial virulence and symbiotic nitrogen fixation. In: Verma DPS (ed) Molecular signals in plant-microbe communications. CRC Press Inc, Boca Raton

Franza T, Expert D (2010) Iron uptake in soft rot Erwinia. In: Cornelis P, Andrews SC (eds) Iron uptake in microorganisms. Horizon Press, Norfolk

Gkouvatsos K, Papanikolaou G, Pantopoulos K (2011) Regulation of iron transport and the role of transferring. Biochim Biophys Acta. doi:10.1016/j.bbagen.2011.10.013

Hantke K (2001) Iron and metal regulation in bacteria. Curr Opin Microbiol 4:172–177

Haas D, Défago G (2005) Biological control of soil-borne pathogens by fluorescent pseudomonads. Nat Rev Microbiol 3:307–319

Hentze MW, Muckenthaler MU, Galy B, Camaschella C (2010) Two to tango: regulation of mammalian iron metabolism. Cell 142:24–38

Huber C, Wachtershaüser G (1997) Activated acetic acid by carbon fixation on (Fe, Ni) S under primordial conditions. Science 276:245–247

Ivanov R, Brumbarova T, Bauer P (2012) Fitting into the harsh reality: regulation of iron-deficiency responses in dicotyledonous plants. Mol Plant 5:27–42

Kaplan CD, Kaplan J (2009) Iron acquisition and transcriptional regulation. Chem Rev 109:4536–4552

Kloepper JW, Leong J, Teintze M, Schroth MN (1980) Enhancing plant growth by siderophores produced by plant growth-promoting rhizobacteria. Nature 286:885–886

Kobayashi T, Nakanishi H, Nishizawa K (2010) Recent insights into iron homeostasis and their application in graminaceous crops. Proc Jpn Acad Ser 86:900–913

Kornitzer D (2009) Fungal mechanisms for host iron acquisition. Curr Op Microbiol 12:377–383
Krewulak KD, Vogel HJ (2008) Structural biology of bacterial iron uptake. Biochim Biophys Acta 1778:1781–1804
Labbé S, Pelletier B, Mercier A (2007) Iron homeostasis in the fission yeast *Schizosaccharomyces pombe*. Biometals 20:523–537
Lemanceau P, Alabouvette C (1993) Suppression of fusarium wilt by fluorescent pseudomonads mechanisms and applications. Biocontrol Sci Technol 3:219–234
Leong SA, Neilands JB (1982) Siderophore production by phytopathogenic microbial species. Arch Biochem Biophys 218:351–359
Lindsay WL, Schwab AP (1982) The chemistry of iron in soils and its availability to plants. J Plant Nutr 5:821–840
Litwin CM, Calderwood SB (1993) Role of iron in regulation of virulence genes. Clin Microbiol Rev 6:137–149
Loeppert RH, Wei LC, Ocumpaugh WR (1994) Soil factors influencing the mobilization of iron in calcareous soils. In: Manthey JA, Crowley DE, Luster DG (eds) Biochemistry of metal micronutrients in the rhizosphere. CRC Press, Boca Raton, pp 343–355
Morrissey J, Guerinot ML (2009) Iron uptake and transport in plants: the good, the bad, and the ionome. Chem Rev 109:4553–4567
Nairz M, Schroll A, Sonnweber T, Weiss G (2010) The struggle for iron—a metal at the host-pathogen interface. Cell Microbiol 12:1691–1702
Neilands JB (1991) A brief history of iron metabolism. Biol Metals 4:1–6
Neilands JB (1995) Siderophores: structure and function of microbial iron transport compounds. J Biol Chem 270:26723–26726
O'Brian MR, Fabiano H (2010) Mechanisms and regulation of iron homeostasis in the rhizobia. In: Cornelis P, Andrews SC (eds) Iron uptake in microorganisms. Horizon Press, Norfolk
Ong ST, Ho JZS, Ho B, Ding JL (2006) Iron-withholding strategy in innate immunity. Immunobiology 211:295–314
Payne SM (1989) Iron and virulence in shigella. Mol Microbiol 3:1301–1306
Raymond J, Segrè D (2006) The effect of oxygen on biochemical networks and the evolution of complex life. Science 311:1764–1767
Robin A, Mazurier S, Mougel C, Vansuyt G, Corberand T, Meyer JM, Lemanceau P (2007) Diversity of root-associated fluorescent pseudomonads as affected by ferritin overexpression in tobacco. Environ Microbiol 9:1724–1737
Rudolph G, Hennecke H, Fischer HM (2006) Beyond the Fur paradigm: iron-controlled gene expression in rhizobia. FEMS Microbiol Rev 30:631–648
Schade AL, Caroline L (1944) Raw egg white and the role of iron in growth inhibition of *Shigella dysenteriae, Staphylococcus aureus, Escherichia coli* and *Saccharomyces cerevisiae*. Science 100:14
Schaible UE, Kaufmann SH (2004) Iron and microbial infection. Nat Rev Microbiol 2:946–953
Singer E, Emerson D, Webb EA, Barco RA, Kuenen JG, Nelson WC, Chan CS, Comolli LR, Ferriera S, Johnson J, Heidelberg JF, Edwards KJ (2011) Mariprofundus ferrooxydans PV-1 the first genome of a marine Fe(II) oxidizing Zetaproteobacterium. PLoS One 6:e25386
Straub KL, Benz M, Schink B (2001) Iron metabolism in anoxic environments at near neutral pH. FEMS Microbiol Ecol 34:181–186
Theil EC (2003) Ferritin protein nanocages use ion channels, catalytic sites, and nucleation channels to manage iron/oxygen chemistry. Curr Opin Chem Biol 15:304–311
Wachtershaüser G (1990) Evolution of the first metabolic cycles. Proc Natl Acad Sci 87:200–204
Weinberg ED (2009) Iron availability and infection. Biochim Biophys Acta 1790:600–605
Widdel F, Schnell S, Heising S, Ehrenreich A, Assmus B, Schink B (1993) Ferrous iron oxidation by anoxygenic phototrophic bacteria. Nature 362:834–836
Williams RJP, Frausto da Silva JJR (1999) Bringing chemistry to life. Oxford University Press Inc, New York
Xu XM, Møller SG (2011) Iron-sulfur clusters: biogenesis, molecular mechanisms, and their functional significance. Antioxid Redox Signal 15:271–307

Chapter 2
Iron in Plant–Pathogen Interactions

Dominique Expert, Thierry Franza and Alia Dellagi

Abstract Infectious diseases are the result of competitive relationships between a host organism and a pathogen. In host vertebrate–microbe interactions, the acquisition of iron for the essential metabolism of pathogenic organisms and the need of the host to bind and sequestrate the metal are central issues. Plants are also confronted with a wide variety of pathogenic microorganisms that can be highly devastating and compromise crop production. Investigated in a few cases in the past, the mechanisms involved in exchanging and withholding iron during plant–microbe interactions are becoming an emerging topic. This chapter surveys the wealth of information illustrating the role of iron acquisition, toxicity, and homeostasis in relevant pathosystem models of agricultural importance. There is now evidence that phytopathogenic bacteria and fungi can use siderophores and other iron uptake systems to multiply in the host and to promote infection. Moreover, plant can develop an iron-withholding response that changes iron distribution and trafficking during infection. Elucidating the mechanisms of competition for iron between plants and pathogens must help to develop innovative strategies for controlling diseases.

D. Expert (✉)
Centre National de la Recherche Scientifique, Laboratoire Interactions
Plantes-Pathogènes, AgroParisTech, 16 rue Claude Bernard, 75005 Paris, France
e-mail: expert@agroparistech.fr

T. Franza
Institut National de la Recherche Agronomique, Laboratoire Interactions
Plantes-Pathogènes, AgroParisTech, 16 rue Claude Bernard, 75005 Paris, France
e-mail: franza@agroparistech.fr

A. Dellagi
Département Sciences de la Vie et Santé, Laboratoire Interactions
Plantes-Pathogènes, AgroParisTech, 16 rue Claude Bernard, 75005 Paris, France
e-mail: dellagi@agroparistech.fr

D. Expert and M. R. O'Brian (eds.), *Molecular Aspects of Iron Metabolism in Pathogenic and Symbiotic Plant–Microbe Associations*, SpringerBriefs in Biometals,
DOI: 10.1007/978-94-007-5267-2_2, © The Author(s) 2012

Keywords Iron acquisition · Iron homeostasis · Oxidative stress · Phytopathogenicity · Plant defense

Infectious diseases are the result of the interaction between a host organism and a pathogen. For the pathogen, entering the host and attaching to target tissues are issues that are tightly controlled by the production of virulence factors which promote the establishment of the microbe and the evasion of host defense. In the later stages of infection, pathogens may have the potential to proliferate beyond their infection sites, causing systemic diseases. In plants, the presence of rigid walls surrounding cells limits the possibilities for the microbe to thrive intracellularly. Plant pathogens produce cell wall degrading enzymes and/or type III secretion systems leading to cell injury and access to nutrients. They encounter diverse environmental conditions in the plant that depend on their colonization sites and their mode of attack. Concerning iron, the availability of this metal for the pathogen may be quite low (see Fig. 2.2). Indeed, plants acquire iron from the environment by the roots using mechanisms based on reduction or chelation (Morrissey and Guerinot 2009). Iron is then mobilized from the root tissues in xylem vessels by citrate which is involved in long distance transport from roots to shoots. Another molecule participating in distribution of iron throughout the plant via the phloem is the nonproteogenic amino acid nicotianamine that is able to chelate iron and other metals (Curie et al. 2009). Iron storage *in planta* occurs at the subcellular level in chloroplasts, where the photosynthetic process takes place (Briat et al. 2010). Plastids contain ferritin, an important iron reservoir protein (Nouet et al. 2011). Vacuoles are other crucial compartments for iron storage and sequestration in plant cells (Lanquar et al. 2005). At the tissue level, the apoplast which includes cell walls and extracellular spaces represent a major site of infection, but the status of iron is not well characterized.

In this context, how do microbes and plants control their iron homeostasis when they are mutually challenging? This chapter surveys the investigations that, in various plant microbial pathosystems, contribute to shed light on this question.

2.1 Iron and Microbial Virulence

2.1.1 Agrobacterium tumefaciens *and Crown Gall Disease*

In early 1980s, the role of iron well established in mammalian infections was still misunderstood in microbial pathogenesis to plants. Interestingly in 1979, J Neilands and collaborators reported the structure of agrobactin, a new catechol-type siderophore produced by the plant pathogenic bacterium *A. tumefaciens* (Ong et al. 1979), and a number of investigations addressing the role of siderophores in the virulence of plant pathogens came to light. *A. tumefaciens* is a soilborne Gram-negative bacterium which produces tumors in a broad range of plants (for a review,

Pitzchke and Hirt 2010). Galls appear at wound sites on the crowns of plants including tobacco, tomato, and potato. Tumor formation results from the transfer of a DNA region (T-DNA) located on a large plasmid present in virulent strains into plant cells, that integrates into the host chromosomal DNA. Expression of bacterial genes in the host leads to the production of enzymes catalyzing the synthesis of plant hormones responsible for cell proliferation, and enzymes generating new metabolites used by bacteria as carbon and nitrogen sources. This association between the pathogen and host cell requires plant signals from wound sites to activate the bacterial *vir* genes involved in T-DNA transfer and a suppressive mechanism of plant defense. To proliferate, tumors do not require the continuous presence of the bacteria.

Agrobactin was characterized as a threonyl peptide of spermidine acylated with three residues of 2, 3-dihydroxybenzoic acid. The production of agrobactin in iron deplete cultures of *A. tumefaciens* strain B6 was associated with the induction of several outer membrane proteins predicted to be involved in ferric-siderophore transport (Leong and Neilands 1981). Mutants unable to produce or utilize agrobactin as iron source were isolated after chemical mutagenesis; they were still capable of initiating tumors in sunflower plants or on carrot root disks. Interestingly, ferric citrate was shown to enhance the growth of the agrobactin deficient mutants exposed to iron deficiency. The authors concluded that the production of agrobactin *in planta* is not required for infection, but citrate could serve as an alternative iron transporter for *A. tumefaciens* within the host. More recently, a gene cluster called *agbCEBA* identified on the linear chromosome of *A. tumefaciens* strain MAFF301001 (Sonoda et al. 2002) was predicted to encode products showing similarity with enzymes involved in biosynthesis of catechol-type siderophores. This cluster revealed to be essential for bacterial growth and production of catechol compounds in iron-limited conditions. In addition, knowledge of the complete genome sequence from *A. tumefaciens* strain C58 allowed the identification of a new gene cluster encoding the biosynthesis of a metabolite with siderophore activity (Rondon et al. 2004). Designated as ATS, this cluster specifies proteins displaying no similarity with other known siderophore biosynthetic enzymes, suggesting that strain C58 produces a novel siderophore. Mutants disrupted in the ATS gene cluster have lost the capacity to produce this molecule and show impaired growth under iron limitation. However, they were able to induce tumors on carrot slices or on tomato seedlings. This result confirms the earlier observation made with mutants of strain B6, and further supports the idea that the production of siderophores is not essential for *A. tumefaciens* to acquire iron *in planta*. On the hand, a high-affinity uptake pathway dependent on a siderophore could be crucial for this bacterium to acquire iron in the soil, during its saprophytic life. As wounding is a prerequisite for successful infection by *A. tumefaciens,* it may be possible that the bacteria at the colonization sites use some iron substrates released in damaged tissues. Indeed, we noted that the C58 genome sequence can encode a number of iron-related transport routes, including two heme utilization pathways and a homologue of the Yfe system. The latter is an ABC permease predicted to transport diverse ferrous and ferric iron ligands (see Table 2.1 and Fig. 2.1).

Table 2.1   Siderophore independent iron routes in some phytopathogenic bacteria

Feo system:

Dickeya dadantii 3937; *D. chrysanthemi* 1591; *D. zeae* 586; *D. paradisiaca* 703
Ralstonia solanacearum GMI 1000; *R. solanacearum* CFBP2957;
Xanthomonas campestris pv. *campestris* B100; *X. oryzae* pv. *oryzae*;
X. oryzae pv. oryzicola; *Xylella fastidiosa*

EfeUOB system:

Dickeya dadantii 3937; *D. chrysanthemi* 1591; *D. zeae* 586; *D. paradisiaca* 703
Erwinia amylovora 1430; *Pseudomonas syringae* pv. *tomato* DC3000; *P. syringae* pv.
syringae B728a; *P. syringae* pv. *tabaci* ATCC 11528, *P. syringae* pv. *phaseolicola*

FecABCDE system:

Pectobacterium atrosepticum SCRI 1043; *P. brasiliensis* 1692; *P. carotovorum* PC1;
Pseudomonas syringae pv. *syringae* B728a; *P. syringae* pv. *tomato* DC3000
P. syringae pv. *tabaci* ATCC 11528; *P. syringae* pv. *phaseolicola*

Hemophore system:

Pectobacterium atrosepticum SCRI 1043; *P. brasiliensis* 1692; *P. carotovorum* PC1 and
WPP14; *P. wasabiae* WPP163; *Agrobacterium tumefaciens* C58

Hmu system:

Dickeya dadantii 3937; *D. chrysanthemi* 1591; *D. paradisiaca* 703; *Pectobacterium
atrosepticum* SCRI 1043; *P. carotovorum* PC1 and WPP14; *P. brasiliensis* 1692; *P.
wasabiae* WPP163; *Agrobacterium tumefaciens* C58

Yfe system:

Dickeya dadantii 3937, *D. chrysanthemi* 1591; *Pectobacterium atrosepticum*
SCRI 1043; *P. carotovorum* PC1 and WPP14; *P. brasiliensis* 1692; *P. wasabiae*
WPP163; *Erwinia amylovora* 1430; *Agrobacterium tumefaciens* C58

Analysis of the *A. tumefaciens* genome revealed the existence of three regulatory genes encoding a homolog of Fur repressor and two proteins, RirA and Irr, originally discovered in rhizobial species (see Chap.3, O'Brian and Fabiano 2010). Kitphati et al. (2007) constructed a *fur* null mutant. These authors showed that in *A. tumefaciens* strain NTL4, the *fur* gene is not involved in iron regulation of siderophore production and transport, but is important for the bacterial survival under iron limiting conditions. Like in *Sinorhizobium meliloti*, *A. tumefaciens fur* gene appeared to negatively regulate the expression of the *sit* operon encoding a manganese uptake system in response to manganese and iron. The *fur* mutant displays hypersensitivity to hydrogen peroxide, low catalase activity, and reduced ability to cause tumors on tobacco leaves compared to wild-type, indicating that the bacteria must accurately control their metal homeostasis and oxidative stress response in early steps of pathogenesis. The Fe–S protein RirA is the rhizobial iron regulator repressing many iron-responsive genes under iron repleted conditions. RirA homologues were found exclusively in members of α-Proteobacteria. A *rirA* null mutation in *A. tumefaciens* led to overexpression of siderophore biosynthesis and transport genes, iron overload, and increased sensitivity to oxidants (Ngok-Ngam et al. 2009). In addition, a functional *rirA* gene appeared to be required for full expression of several *vir* genes induced by the plant compound

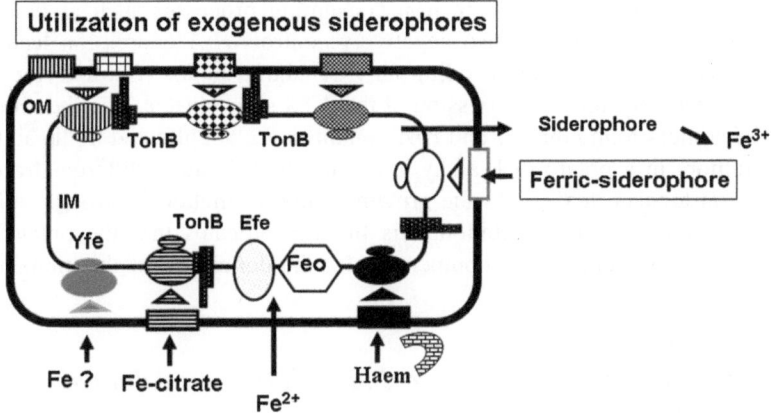

Fig. 2.1 Schematic representation of the various iron uptake routes in phytopathogenic Gram-negative bacteria (for a review see Crosa et al. 2004). Bacterial cells can synthesize and excrete siderophores that form with Fe^{3+} a siderophore ferric complex, designated as ferric-siderophore. Metal chelating power of siderophores are compared by calculating pFe values (pFe $= -\log$ [Fe^{3+}] at a defined concentration and pH). A ferric-siderophore is specifically recognized by an outer membrane TonB-dependent transporter (*rectangles*), which is a gated-channel energized by the cytoplasmic membrane-generated proton motive force transduced by the TonB protein and its auxiliary proteins ExbB and ExbD. Transport of a ferric-siderophore across the inner membrane involves a less specific ABC permease (*triangle and circles*). Ferric complexes of citrate or of exogenous siderophores, i.e., produced by other microbes are imported in a similar way. Heme can be transported in two ways: either directly or bound to a secreted hemophore protein which delivers it in the cytoplasm, via specific TonB-dependent transporters and specific ABC permeases. Ferrous iron can be transported through the FeoAB and/or EfeUOB systems. FeoB, the main component of the Feo system, is an integral membrane protein with an N-terminal domain having GTPase activity essential for the transport function. Function of the FeoA peptide is unknown. EfeU is a potential integral inner protein acting as a permease for ferrous or ferric forms of iron. Functions of EfeO and EfeB proteins are unknown. The Yfe ABC permease can import an uncharacterized form of iron. The nature of the Fe-nicotianamine uptake system is unknown. OM: outer membrane, IM: inner membrane

acetosyringone, and for tumor formation on tobacco leaves. Thus, the RirA protein seems to play an important role in the control of *A. tumefaciens* pathogenicity. In *Bradyrhizobium japonicum*, the Irr protein controls iron homeostasis through heme biosynthesis. Heme biosynthesis is an iron-dependent process, the final step residing in insertion of iron into protoporphyrin, a reaction catalyzed by ferrochelatase. In the absence of iron, Irr represses heme biosynthetic genes and activates iron transport genes. In the presence of iron, Irr interacts with ferrochelatase, leading to Irr degradation via heme binding and derepression of the heme biosynthetic pathway (see Chap.3). The role of *A. tumefaciens irr* gene in iron metabolism was investigated. By constructing deletion mutations in *irr* and *rirA*, alone and in combination, Hibbing and Fuqua (2011) showed that these two genes work in a complementary way to maintain a tight control of iron homeostasis. Under conditions of iron limitation, the *irr* gene negatively controls the transcription of genes involved in iron consuming processes such as heme and Fe–S cluster biosynthesis. It also negatively controls *rirA* expression leading to derepression of *sigI*, a gene encoding an ECF sigma factor (Extra Cytoplasmic Function) necessary for the expression of the ATS gene cluster described in strain C58. The authors found that *irr* and *rirA* mutants were not affected in the ability to induce tumors in a potato disk assay, a result which is quite different from that observed on tobacco leaves. As the growth of the *irr* mutant is strongly affected under conditions of iron limitation, this finding indicates that the potato disks might supply sufficient iron to compensate for the deficiency of this mutant.

2.1.2 Pseudomonas syringae, *a Plant Pathogen Adapted to Different Hosts*

Pseudomonads are widespread Gram-negative bacteria displaying various lifestyles and habitats. *P. aeruginosa*, an opportunistic pathogen infecting humans and animals is the most investigated species; *P. fluorescens/putida* and *P. syringae* are well known for their ecological importance in soils and implication in plant disease, respectively (for a review Silby et al. 2011). *P. syringae* comprises numerous pathovars causing symptoms that may vary according to the host species and the site of infection. A typical phenotypic trait of Pseudomonads is the production in iron deficient environments of green-yellow fluorescent compounds emblematic of a class of siderophores called pyoverdines. Pyoverdines contain a conserved dihydroxyquinoline chromophore linked to a variable peptide chain considered as a taxonomic hallmark (Meyer et al. 2008). Their high affinity for the ferric ion is conferred by the presence of catecholate and hydroxamate groups. Pyoverdine contributes to iron acquisition by *P. aeruginosa* in vivo. This class of siderophores also plays an important role in the biological control of phytopathogenic microrganisms of the rhizosphere and on the iron availability to plants in soils (Visca et al. 2002).

İnterestingly, by investigating various pathovars of *P. syringae* for the production of phytotoxins, D Gross and collaborators found that a maximum yield of the necrosis inducing toxin syringomycin was obtained, if micromolar ferric iron concentrations were supplied to the growth medium (Gross 1985). This fact suggested that the pathogen could necessitate relatively high concentrations of iron during infection, and characterization of the fluorescent pigment of *P. syringae* pv. *syringae* was undertaken (Cody and Gross 1987). A pyoverdine-like molecule that confers on the bacterium the ability to grow under iron limitation was identified, and pyoverdine biosynthetic and transport mutants were isolated; these mutants were able to cause necrotic lesions on sweet cherry fruit equivalent to those observed with the wild-type strain indicating the dispensability of pyoverdine *in planta*. However, with emergence of complete genomic sequences of various *P. syringae* strains, it became apparent that most of pathovars not only have the capacity to produce pyoverdines, but they also could synthesize pyoverdine unrelated siderophores. The genome of *P. syringae* pv. *tomato* DC3000 revealed to contain a gene showing similarity with the polyketide synthase gene *irp1* present in the YBT yersiniabactin gene cluster of *Yersinia pestis.* Yersiniabactin is a salicylate derived siderophore widespread among human and animal pathogenic bacteria and contributes to their virulence (Perry and Fetherston 2011). In *Yersinia* spp, the YBT locus is located in a genomic high-pathogenicity island. In *P. syringae,* Bultreys et al. (2006) reported the presence of an YBT locus in pathovars *tomato* and *phaseolicola*, the gene organization of which appeared to be well conserved through the different strains examined. In agreement with gene predictions, yersiniabactin was detected in the culture medium of the majority of strains tested. However, the occurrence of yersiniabactin nonproducers among natural isolates allowed the authors to consider that this siderophore is not essential for the pathogenicity in *P. syringae.*

The role of siderophores was further explored in the model strain *P. syringae* pv. *tomato* DC3000 that causes bacterial speck of tomato and is also virulent on *Arabidopsis.* Jones et al. (2007) found that this strain produces salicylic acid, yersiniabactin, and pyoverdine, in response to iron limitation. These authors showed that salicylic acid is a precursor for yersiniabactin synthesis and that formation of yersiniabactin is dependent on the isochorismate synthase gene *pchA.* Using a *pchA* insertion mutant, they investigated the physiological importance of yersiniabactin. Yersiniabactin appeared to be produced by the wild-type strain but not the *pchA* mutant during infection of *Arabidopsis* leaves. However, the *pchA* mutation had no effect on the ability of bacteria to colonize plant leaves, to multiply *in planta,* and to produce symptoms. The role of pyoverdine in pathogenesis was also studied (Jones and Wildermuth 2011). The loss of pyoverdine production in a *pvd* insertion mutant impaired in the pyoverdine biosynthetic pathway, had no impact on the virulence of DC3000 on tomato plants. Using a double mutant unable to synthesize yersiniabactin and pyoverdine, the authors showed that strain DC3000 produces a third siderophore-like molecule identified as citrate. A yersiniabactin and pyoverdine negative mutant missing the ferric citrate transporter FecA displayed a severe growth defect in iron-limited culture but was fully pathogenic, suggesting that DC3000 could obtain iron *in planta* by other means. Indeed, the genome of DC3000 reveals the presence of a homologue of the Yfe permease which might be effective during infection. An interesting hypothesis proposed by the authors is that during infection by DC3000 the leaf apoplast would not be limited in iron. In this regard, it is noteworthy that Kim et al. (2009, 2010) found that high concentrations of iron in the culture medium repressing expression of high-affinity iron transport systems can induce expression of the type III secretion system and virulence genes. Thus, there would be sufficient amounts of iron at the sites of infection by DC3000 to trigger the pathogenic process. A study of the transcriptional profile of *P. syringae* pv. *phaseolicola* NPS3121 in response to tissue extracts from a susceptible pea cultivar may also support this hypothesis, since most repressed genes were found to concern the uptake and metabolism of iron (Hernandez-Morales et al. 2009).

Contrastingly, Taguchi et al. (2010) reported the importance of pyoverdine as a virulence factor of *P. syringae* pv. *tabaci* 6605 in tobacco. The authors identified a pyoverdine gene cluster and constructed deletion mutants defective in biosynthesis of the pyoverdine peptide chain or the chromophore, as well as a mutant missing the ferripyoverdine receptor. On tobacco leaves, the pyoverdine-deficient mutants were less virulent than the wild-type and their growth *in planta* was reduced. Genetic complementation of the mutations restored pyoverdine production and virulence properties. However, the virulence of the receptor mutant was not affected. Further characterization of these mutants revealed that all have acquired lower abilities to produce tabtoxin, extracellular polysaccharide, and quorum-sensing molecules, as well as increased swarming ability and surfactant production. In light of the regulatory role conferred by pyoverdine and its receptor in *P. aeruginosa* PAO1 (Visca et al. 2002), it was proposed that in *P. syringae* pv. *tabaci* 6605, pyoverdine also acts as a signal triggering the production of several pathogenicity factors. Indeed, *P. aeruginosa* pyoverdine and its receptor mediate a transduction signal inducing the expression of ECF factors PvdS and FpvI that control a set of functions of which iron acquisition and pathogenicity. A *pvdS* gene orthologue was identified in the DC3000 genome. On the basis of a computational analysis, Swingle et al. (2008) characterized the DC3000 PvdS regulon and the PvdS-regulated promoter motif. They concluded that PvdS is a well conserved regulator of pyoverdine synthesis among fluorescent pseudomonads, although the promoters recognized by this factor are likely to differ from species to species. A homologue of the *E. coli* ECF sigma factor FecI involved in ferric citrate transport and signaling was also identified in DC3000. The DC3000 FecI protein was shown to regulate gene expression through a signal transduction cascade mediated by a TonB-dependent receptor in response to exogenous siderophores (Markel et al. 2011). In addition, different *P. syringae* pathovars appeared to contain a homologue of the *fur* gene (Butcher et al. 2011). By constructing a *fur* deletion mutant in *P. syringae* pv *tabaci* 11528, Cha et al. (2008) demonstrated that *fur* not only regulates siderophore production, but it also controls swarming motility as well as the production of tabtoxin and quorum-sensing molecules. The virulence of the *fur* mutant was compromised as revealed by the reduced size of necrotic lesions, the decline of bacterial populations and the absence of chlorosis around the lesions in infected tobacco leaves. These data indicate the importance of this iron sensory regulatory gene in the interaction between *P. syringae* pv *tabaci* and its host plant.

Later, it was found that many strains of the *P. syringae* group are able to secrete a second siderophore, achromobactin, structurally different from yersiniabactin. Achromobactin is a citrate/carboxylate siderophore originally described in the bacterial pathogen *Dickeya dadantii* (Münzinger et al. 2000, see Sect. 6.a). Among the six *P. syringae* pathovars the genome sequences of which could be consulted, only *P. syringae* pv. *tomato* DC3000 does not possess an achromobactin biosynthesis gene cluster. Mechanistically, achromobactin biosynthesis was reported as an excellent model for studying siderophore synthases having a mode of action different from that of nonribosomal peptide synthase (NRPS) (Schmelz et al. 2008, 2011). Indeed a major biosynthesis pathway for siderophores relies on the extensively investigated NRPS multi-enzyme superfamily. Interestingly, an analysis of the synthetases AcsA, AcsC, and AcsD in *P. syringae* pv. *syringae* B728A allowed the in vitro reconstitution of achromobactin biosynthesis (Berti and Thomas 2009). Owen and Ackerley (2011) investigated the role of pyoverdine and achromobactin in the pathogenicity of *P. syringae* pv. *phaseolicola* 1448A. They tested the ability of siderophore negative mutants to form lesions on bean pods but neither pyoverdine nor achromobactin proved to be essential for this bacterium to cause halo blight. The authors suggested that the pathogenicity assay used in their study was unlikely the most appropriate to appreciate differences in disease progression, since lesion formation depends primarily on toxin production.

Interestingly, *P. syringae* comprises a number of strains exhibiting a pronounced epiphytic phase on plants during which they can establish high population levels on the leaf surface (Feil et al. 2005). Epiphytic populations of *P. syringae* pv. *syringae* B728A can subsequently invade the apoplastic spaces of plant tissue and initiate disease. On the other hand, there are natural isolates which are nonpathogenic and can behave as microbial antagonists able to suppress diseases induced by pathogenic strains. For example, *P. syringae* pv. *syringae* 22d/93 exerts a biocontrol activity against *P. syringae* pv. *glycinea*, the causal agent of bacterial blight of soybean (Wensing et al. 2010). Such epiphytes produce the siderophores, pyoverdine, and achromobactin. Disruption of either pyoverdine or achromobactin biosynthesis impairs the epiphytic fitness of strain 22d/93, as observed by a significant decrease in the population size of the mutants on soybean leaves compared to wild-type. The growth deficiency of these mutants was compensated for when wound inoculation was used, indicating the availability of iron in the presence of small lesions on the leaves. Thus, rather than serving as a virulence factor, production of achromobactin and pyoverdine by *P. syringae* pv. *syringae* 22d/93 contributes to the epiphytic fitness of this bacterium and improves its competitiveness as an antagonist of virulent strains. In *P. syringae* pv. *syringae* B728A, expression of achromobactin mediated iron transport was shown to depend on a gene cluster which also specifies the production of a new ECF sigma factor, AcsS, regulating the biosynthesis and secretion of achromobactin, as well as, other genes associated with epiphytic growth and survival (Greenwald et al. 2012). A complex iron regulatory network must exist in B728A, and further studies are needed to better define the role of siderophores on the leaf surface.

2.1.3 *Diseases Caused by* Xanthomonas *Species*

Plant pathogenic *Xanthomonas* spp. form a large group of Gram-negative bacteria which cause various diseases like vascular wilts, cankers, leaf spots, fruit spots, and blights on many plants and economically important crops (Ryan et al. 2011). These bacteria initially grow on leaf surfaces and enter into the host through hydathodes or wounds to spread systemically through the xylem elements of the vascular system or through stomata to colonize the mesophyll parenchyma.

X. campestris includes host-specific pathovars that infect different brassicaceous, solanaceous, and other plant species, whereas *X. oryzae* infects rice and some of its wild relatives.

Despite the availability of genome sequences of many *Xanthomonas* strains showing the existence of siderophore biosynthetic gene clusters in different pathovars, the structure of the siderophores produced is still unknown (Etchegaray 2004, Pandey and Sonti 2010). In this respect, the report published by Wiggerich and Puhler (2000) illustrating the difficulty to elucidate the role of a TonB-dependent ferric iron acquisition system in the pathogenicity of *X. campestris* pv. *campestris* B100 is interesting. These authors found that a TonB negative mutant, thus deficient in ferric iron ligand uptake did not induce the typical phenotype of black rot symptoms on leaf of cauliflower, but spread systemically in this plant, suggesting that this mutant could meet its iron needs *in planta* via a mechanism independent of the TonB machinery. However, this mutant neither grew and nor induced a hypersensitive response on pepper, a nonhost plant. These intriguing results remained unexplained. It may be assumed that the inability of the *tonB* mutant to cause black rot symptoms was due to a pleiotropic effect of the mutation thus altering a function unrelated to iron transport. In a mutant screening to find out new virulence functions of *X. oryzae* pv. *oryzae*, Chatterjee and Sonti (2002) isolated a strain deficient in the ability to produce lesions on a susceptible rice cultivar. Molecular characterization revealed that this mutant was impaired in a gene homologue to the *X. campestris* pv. *campestris rpfF* gene required for production of a diffusible extracellular factor that positively regulates virulence associated functions. In *X. oryzae* pv. *oryzae*, the virulence deficiency of the *rpfF* mutant was corrected by iron supplementation. This mutant overproduced siderophores and was sensitive to iron deficiency, a phenotype that could also be corrected by iron supplementation. Thus, iron metabolism plays a critical role in *X. oryzae* pv. *oryzae* virulence. Homologues of the *feoABC* genes that encode a ferrous iron uptake system called Feo and characterized in *E. coli* (Fig. 2.1) were identified in this pathovar (Pandey and Sonti 2010). A *feoB* mutation predicted to impair the production of the main component of ferrous iron transport caused severe defects in *X. oryzae* pv. *oryzae* growth under iron-limited conditions and virulence. In the other hand, mutations in a gene cluster encoding proteins involved in biosynthesis and utilization of the *X. oryzae* pv. *oryzae* siderophore, the *xss* operon, had no impact on the virulence. Using histochemical and fluorometric assays with *uidA* gene reporter fusions, the authors demonstrated that during infection of the susceptible rice cultivar, the bacteria express the *feoB* gene while the *xss* operon remains turned off. *X. oryzae* pv. *oryzae* grows in the rice xylem vessels and in this regard, it is worth noting that the iron content in the xylem sap of rice seedlings cultured under iron proficient conditions reaches a concentration range sufficient to satisfy the bacterial iron needs. In this fluid, the ferrous form of iron predominates while the ferric form is mainly present as ferric citrate (Yokosho et al. 2009). *X. oryzae* pv. *oryzae* genome analysis revealed the absence of a potential ferric citrate transport system, thus making particularly relevant the fact that the *feoB* mutant displays attenuated virulence. The role of the

Fur regulatory protein in pathogenicity of *Xanthomonas* spp was also investigated (Subramoni and Sonti 2005; Jittawuttipoka et al. 2010). First characterized in *X. campestris* pv. *phaseoli*, the *fur* gene was found to be well conserved among *Xanthomonas* spp, although it encodes a protein with unusual features (Loprasert 1999). The cysteine residues highly conserved at the Zn binding motifs are missing. In *X. oryzae* pv. *oryzae* as well as in *X. campestris* pv *campestris*, analysis of a *fur* mutant indicated that the Fur protein regulates the production of sidero-phores. In absence of a functional *fur* gene the bacterial cells are hypersensitive to oxidative stress, accumulate reactive iron species and have decreased catalase activity as well as aerobic growth defects. The *fur* mutants showed attenuated virulence properties on their respective hosts, rice, and chinese cabbage, as visu-alized by a decreased lesion size on the leaves. The growth deficiency in rice leaves of the *fur* mutant of *X. oryzae* pv. *oryzae* appeared to be rescued by application of an antioxidant such as ascorbic acid, indicating that this growth defect was likely due, in part to an impaired ability of the bacteria to cope with the oxidative stress conditions encountered during infection.

2.1.4 Diseases Caused by Xylella fastidiosa

X. fastidiosa is a Gram-negative bacterium associated with diseases of economical importance like the Pierce's disease of grape and the variegated chlorosis of citrus species (Chatterjee et al. 2008). This pathogen is vector-transmitted from one plant to another by various xylem sap-feeding insects. Symptoms are typically a leaf-scorch associated with the extensive colonization of xylem vessels. From infection sites, bacterial cells attach to the vessels walls and multiply forming biofilm-like colonies that can completely occlude xylem vessels, thereby blocking water transport. *X. fastidiosa* is closely related to various *Xanthomonas* species and the availability of complete genome sequences for both genera allowed the discovery that many genes implicated in the virulence of *Xanthomonas* species have homologues in *X. fastidiosa*. However, genes encoding type III secretion system machinery and associated effectors are missing in *X. fastidiosa*. Zaini et al. (2008) were interested in investigating the response to changes in iron availability of a strain pathogenic to citrus. By analyzing the transcriptional profile of *X. fastidiosa* strain 9a5c exposed to large variations of iron levels, these authors identified an iron stimulon encompassing the genes involved in uptake and storage of this metal as well as those responsible for type IV pilus and colicin V production. The type IV pilus contribute to twitching mediated motility and mutants deficient in type IV pili in the host are inhibited from colonizing upstream vascular regions. Thus, an attractive hypothesis is that iron could be sensed as an environmental signal susceptible to modulate the transcriptional control of this virulence factor. Investigations of the contribution of Fur protein in the regulation of this iron stimulon suggest that additional transcriptional regulators are involved in the response to iron. Of the genes participating in iron transport and up-regulated under low iron conditions, those encoding the Feo system and several TonB-dependent receptors would be interesting to investigate as pathogenicity determinants.

2.1.5 Ralstonia solanacearum, *The Causal Agent of Bacterial Wilt on a Large set of Plants*

R. solanacearum is a soilborne Gram-negative bacterium that causes lethal wilt diseases of many plants including economically important crops around the world (Genin 2010). The bacterium enters plant roots through natural or mechanical wounds, multiplies in the root cortex and bacterial invasion of xylem vessels leads to systemic spreading. Wilting symptoms occur as a consequence of an intensive multiplication of the bacteria and are associated with the large production of exopolysaccharides blocking water traffic in the plant. Though a wide range of genes and functions playing a role in the colonization and multiplication of the bacterium within plant tissues are characterized, the way by which *R. solanacearum* meets its iron requirements is less well documented. It was interesting to discover that the *phcA* gene encoding a transcriptional regulator belonging to a cell density sensing system essential for production of *R. solanacearum* virulence factors is located immediately upstream of the *fur* gene on the genome of *R. solanacearum* strain AW1. Intrigued by the juxtaposition of these two sensory regulatory genes, Bhatt and Denny (2004) investigated the production of sidero-phores in this bacterium. They found that under iron limitation *R. solanacearum* AW1 releases a siderophore of the carboxylate family structurally determined as staphyloferrin B also produced by *R. metallidurans* and *Staphylococcus aureus*. The *phcA* gene was shown to negatively regulate the production of this sidero-phore and several pathogenicity factors. The authors isolated insertion mutants with reduced siderophore activity and identified a locus predicted to be involved in biosynthesis, export, and transport functions of staphyloferrin B. These mutants revealed to be fully pathogenic after inoculation on unwounded roots of young tomato plants. There was no production of siderophore in a culture of the wild-type strain grown in xylem sap from tomato plants. These data indicate that *R. solanacearum* AW1 acquires iron in the xylem vessels by a transport route different from that mediated by staphyloferrin B. The availability of the genome sequence of *R. solanacearum* GMI1000 led to the finding that this strain produces a metal-complexing antibiotic, micacocidin previously identified as a product structurally related to the siderophore yersiniabactin (Kreutzer et al. 2011). This molecule is synthesized by a number of staphyloferrin B producing strains, and it would be of interest to know whether it acts as a siderophore.

An in vivo expression technology (IVET)-like screen performed in *R. solanacearum* strain UW551 identified up-regulated genes in the presence of tomato root extracts (Colburn-Clifford et al. 2010). One of these genes, *dps*, encodes a miniferritin initially defined in *E. coli* as a DNA-binding protein from starved cells and found to protect DNA from oxidative damages by interacting with DNA. However, Dps proteins possess a ferritin-like function that endows them with iron and hydrogen peroxide detoxification properties (for a review see Chiancone and Ceci 2010). Dps proteins are also called miniferritins, because they are assembled from only 12 identical subunits rather than the 'canonical' 24 subunits. It was demonstrated that expression of the *dps* gene of *R. solanacearum* contributes to oxidative stress tolerance and to colonization of tomato plants. However, the role that this protein may play in iron metabolism of this species has not been explored.

2.1.6 Enterobacterial Species and Plant Disease

In the past, the *Erwinia* genus was the only member of the *Enterobacteriaceae* family representative of the plant kingdom, including mostly phytopathogenic species. Exposed to many nomenclatural difficulties, the various pectinolytic species/subspecies classified in this genus were divided in three new genera, *Pectobacterium*, *Dickeya*, and *Brenneria*. These species are broad-host range pathogens causing soft rot diseases to economically important plants and crops, including vegetables and ornamentals (Czajkowski et al. 2011). Now, the *Erwinia* genus comprises non pectinolytic species displaying limited host range. The most important one, *E. amy*lovora, is responsible for fire blight to apple, pear, and other rosaceous plants (Eatsgate 2000).

2.1.6.1 Pectinolytic Species and Soft rot Disease

In the mid-1980s, a work aimed at investigating the pathogenicity determinants of *D. dadantii* (formerly *E. chrysanthemi*) revealed a possible role of iron assimilation in plant pathogenesis (Expert and Toussaint 1985). *D. dadantii* 3937 produces a systemic disease in African violets as well as in *Arabidopsis*. Symptoms consist of tissue disorganization due to the release of a set of bacterial pectinolytic enzymes that degrade plant cell walls. Cell wall deconstruction weakens plant cells, allows bacteria free access to cellular nutrients and facilitates their dissemination throughout the leaf and petiole. Colonization of leaf tissues begins with a symptom-less phase during which bacterial cells remain clustered in intercellular spaces and then migrate intercellularly without causing severe injury of cellular structures (Murdoch et al. 1999). Bacterial cells do not invade the vascular tissues. Searching for bacteriocin resistant mutants with impaired envelope structure, the authors isolated clones lacking low iron inducible proteins in their outer membrane and unable to cause systemic symptoms in African violets. A role in ferric-siderophore transport was hypothesized for these proteins and thus, the possibility that they contribute to iron nutrition of this pathogen *in planta* was considered. Induction of these proteins was correlated with production of a monocatechol siderophore called chrysobactin (Persmark et al. 1989). Further investigations demonstrated that other strains of soft rot *Erwinia*, such as *E. carotovora* W3C105 (Barnes and Ishimaru 1999) and *D. chrysanthemi* EC16 (Sandy and Butler 2011) can produce chrysobactin when exposed to iron restriction. The genetics of chrysobactin mediated iron transport was explored in depth in *D. dadantii* 3937 (Franza and Expert 1991; Rauscher et al. 2002; Expert et al. 2004). Interestingly, chrysobactin biosynthesis mutants were found to respond positively to the chemical assay used for the detection of siderophores (Schwyn and Neilands 1987), and this led to the discovery of a second iron transport route mediated by achromobactin (Mahé et al. 1995; Münzinger et al. 2000). Chrysobactin-deficient mutants cannot compete with a strong ferric iron chelator but owing to achromobactin they can still thrive on a medium containing a ferrous iron chelator like 2,2' dipyridyl. A genetic cluster encoding all proteins necessary for achromobactin biosynthesis and transport was characterized in strain 3937 (Franza et al. 2005). The genome of related species, *D. chrysanthemi* 1591 and *D. zeae* 586, reveals the existence of a similar gene locus.

Several studies demonstrated that chrysobactin and achromobactin highly contribute to successful infection of the plant (Enard et al. 1988; Franza et al. 2005; Dellagi et al. 2005). Chrysobactin defective mutants produce only localized symptoms on African violets and compared to a chrysobactin proficient strain their growth *in planta* is reduced. This decline in population size appeared to coincide with the emergence of a necrotic border surrounding the lesion (Masclaux and Expert, 1995). Chrysobactin was detected in leaf intercellular fluids from plants inoculated with the wild-type strain, suggesting that this compound could sequester the iron present in colonized tissues and possibly induce a reaction from the plant able to deprive the bacterial cells of essential iron (Neema et al. 1993). By investigating the coordination properties of the different ferric complexes of this siderophore, the Albrecht-Gary's group (Tomisic et al. 2008) found that chrysobactin is a less effective ferric chelator than hexadentate siderophores, such as enterobactin or desferrioxamine B. However, due to a higher pFe value (as defined in Fig. 2.1) than citrate or malate which are major ferric ion carriers in plants (pFe of chrysobactin = 17.1 vs. pFe of citrate = 14.8), chrysobactin can effectively sequester the iron from these plant chelators. Achromobactin deficient mutants are also affected in their virulence but are more aggressive than the chrysobactin nonproducers and double mutants deficient in both achromobactin and chrysobactin production, are impaired in symptom initiation. The wild-type cells but not the double mutants can survive for several days in intercellular spaces of host tissues without multiplying substantially. During the symptomatic phase, only the wild-type cells can proliferate.

The low availability of iron for the bacteria in the apoplast acts as a signal that not only turns on the transcription of genes involved in iron assimilation, but also those encoding the major pectin-degrading enzymes encoded by the *pelD* and *pelE* genes (Franza et al. 1999, 2002). This regulation, mediated by the transcriptional repressor Fur, allows a fine control of the pathogenicity in response to intracellular iron levels. It implies two distinct mechanisms. DNase I footprinting experiments demonstrated that the Fur binding sites covers the -35 and -10 promoter element of the ferric chrysobactin receptor encoding gene *fct* suggesting a direct competition between the RNA polymerase and Fur. On the other hand, for the *pelD* and *pelE* gene promoters, the sequence protected by Fur is located upstream from the -35 promoter element and includes a part of the binding site of the cAMP receptor protein CRP, required for activation of *pel* gene transcription. In this case, Fur would act as an antiactivator of transcription by blocking the action of CRP (Franza et al. 2002). Thus, the two pathogenicity determinants, iron acquisition and production of pectinases are regulated in a coordinated manner and this metabolic coupling can confer an important advantage on *D. dadantii* cells during pathogenesis. When inoculated on African violets, the *fur* mutant displayed an altered virulence in comparison to that of the wild-type strain. This reduced pathogenicity of the *fur* mutant on its host was explained by its altered growth capacity *in planta* (Franza et al. 1999) indicating that a tight control of the bacterial intracellular iron content is necessary for full virulence.

Indeed, during infection bacterial cells have to cope with the production of reactive oxygen species by plant cells (Santos et al. 2001; Fagard et al. 2007). Several studies demonstrated the importance of a regulatory link between iron metabolism and tolerance to oxidative stress. In particular, there was the discovery that the Suf machinery involved in the formation of Fe–S clusters under iron starvation and oxidative conditions is necessary for full virulence (Nachin et al. 2001). Microarray profiling of bacterial genes that are specifically up- or down-regulated in saintpaulia leaves, as well as in vivo expression analysis of promoter–GFP (Green fluorescent protein) reporter constructs in leaves of spinach allowed the identification of *E. chrysanthemi* genes that are regulated during plant infection (Okinaka et al. 2002; Yang et al. 2004). Upregulation of genes involved in iron uptake and in stress responses to reactive oxygen species was observed. *E. chrysanthemi* possesses three ferritins, of which the heme-free ferritin FtnA and the heme-containing bacterioferritin Bfr play differential roles in virulence, depending on the host (Boughammoura et al. 2008). FtnA constitutes the main iron storage protein. Indeed, an *ftnA* mutant that lacks a functional FtnA ferritin has increased sensitivity to iron deficiency compared to the wild-type. In addition, this mutant has increased sensitivity to oxidative and nitrosative stresses, as well as an increased content in ferrous iron. By limiting the concentration of reactive iron, FtnA reduces the cytotoxic effect of the Fenton reaction, and thus confers tolerance to oxidative stress. Bacterioferritin acts as a transient iron store which plays an important role in distribution of iron between essential metalloproteins, particularly under conditions of iron deficiency (Expert et al. 2008). In contrast, the *D. dadantii* miniferritin encoded by the *dps* gene has a minor role in iron homeostasis, but is important in conferring tolerance to hydrogen peroxide and for survival of cells that enter the stationary phase of growth (Boughammoura et al. 2012).

A comparative genomic study of *P. atrosepticum* and *D. dadantii* was carried out (Franza and Expert 2010). It appeared that besides the production and utilization of siderophores, *P. atrosepticum* and *D. dadantii* have the capacity to use other iron sources (Table 2.1). Indeed, both species are able to use heme iron, whereas only *P. atrosepticum* can transport the ferric citrate complex and only *D. dadantii* can acquire ferrous iron. These different modes of iron capture indicate that these species have to cope with various environmental and ecological conditions during their pathogenic cycle. For instance, the *D. dadantii* FeoAB system is likely to be functional during plant infection. Construction of a *D. dadantii feoB* negative mutant allowed to show that when present in an achromobactin and chrysobactin negative background, the *feoB* mutation could confer a three-fold reduction in iron uptake under reducing conditions, compared to the *feoB* positive strain. With respect to the pathogenicity on Arabidopsis plants, there was no significant effect of the *feoB* mutation, when present in a wild-type background. On the other hand, in a siderophore negative background, the *feoB* mutation resulted in a reduced number of systemic infections, which was twofold lower than with the siderophore nonproducer and five times lower than with the wild-type strain (Franza and Expert 2010). Okinaka et al. (2002) also showed that *D. dadantii feoB* gene is up-regulated during infection of African violets.

2.1.6.2 Erwinia amylovora and Fire Blight Disease

Fire blight caused by *E. amylovora* is characterized by a progressive necrosis of tissues of infected aerial parts of the plant and often associated with ooze production. Natural infections mainly occur through wounds and natural openings, especially on flowers. On susceptible hosts, bacteria first move through the intercellular spaces of the parenchyma and at a later stage into the xylem vessels, provoking extensive lesions and sometimes complete dieback of the tree. The strategy of infection of *E. amylovora* differs from that of *D. dadantii* in that the two main pathogenicity factors for this bacterium are the exopolysaccharide amylovoran and secretion of effectors through the type III secretion system. These effectors are required during early infection steps that lead to local necrosis, whereas amylovoran is required during later stages which result in bacterial progression *in planta*. *E. amylovora* effector proteins contribute to pathogenesis by controlling plant defenses and triggering cell death (Degrave et al. 2008).

   In iron-limited environments, *E. amylovora* produces cyclic trihydroxamate siderophores belonging to the desferrioxamine family, with a predominance of deferrioxamine E. First isolated from actinomycetes, desferrioxamines were also found in other bacteria such as *E. herbicola* (Feistner et al. 1993; Kachadourian et al. 1996). Unlike nonpathogenic mutants lacking a functional type three secretion system, mutants affected in ferrioxamine-mediated iron transport still cause fire blight (Dellagi et al. 1998, 1999). The lack of ferrioxamine-dependent iron uptake had no effect on pathogenicity when tested on apple seedlings. However, the mutants were less able to colonize floral tissues and to initiate necrosis on apple flowers, indicating that the production of desferrioxamine is critical at the onset of infection. This observation led the authors to speculate that desferrioxamine is involved not only in iron acquisition, but also in oxidative stress responses of the pathogen. Indeed, *E. amylovora* wild-type strains induce electrolyte leakage from host plant cells as the result of cell death. The ability of desferrioxamine biosynthetic mutants to induce this reaction is severely reduced and this defect is rescued by the addition of this siderophore. As desferrioxamine alone does not induce electrolyte leakage, it was proposed that this compound, by inhibiting the generation of toxic radicals via the Fenton-type redox chemistry, protects the bacterial cells against the toxic effects of reactive oxygen species produced at the onset of infection. The reduced ability of the desferrioxamine biosynthetic mutants to cause electrolyte leakage would result from the transient decrease in bacterial population size following the initial oxidative burst. In support of this interpretation was the increased survival of *E. amylovora* cells treated with hydrogen peroxide in the presence of desferrioxamine B, which is known to protect animal tissues from damage by reactive oxygen species. By forming high-stability iron complexes, compounds of the desferrioxamine family inhibit the generation of hydroxyl radicals via the Fenton reaction. Interestingly, Zhao et al. (2005) found that the gene encoding the ferritin Ftn is induced during infection in pear tissue.

A search of the complete genome of *E. amylovora* 1430 for the presence of the *foxR* gene encoding the ferrioxamine receptor allowed Smits and Duffy (2011) to identify a genetic cluster involved in desferrioxamine biosynthesis and transport. These authors found the existence of a high conservation of the desferrioxamine biosynthetic proteins in genomes of members of the related genera *Erwinia* and *Pantoea* indicating the ancestral nature of this feature.

2.1.7 *Pathogenicity of Ascomycete Fungi*

The filamentous ascomycete *Cochliobolus heterostrophus* is representative of a genus that attacks monocots, including all major cereal crops, worldwide. This fungus was known as a mild pathogen of corn until 1970, when a highly virulent race caused the Southern Corn Leaf Blight (SCLB), which devastated the US corn crop along the eastern seaboard. Thus, *C. heterostrophus* emerged as a model necrotrophic fungus for the study of plant pathogenesis (Turgeon and Baker 2007). The fungus overwinters as conidia and mycelia on debris of dead corn plants. Conidia infect corn leaves by direct penetration and cause small lesions, which, under epidemic conditions, can cover the entire leaf, thereby killing the plant. *C. heterostrophus* and related taxa are known for their ability to produce host-specific toxins that serve as pathogenicity factors. Synthesized by NRPS, each toxin appears to be necessary for development of a particular disease. However, in a number of physiological and genetic studies, they were found to be insufficient to explain pathogenicity entirely. Looking for general pathogenicity factors, Lee et al. (2005) undertook a genome-wide search for NPRS-encoding genes in *C. heterostrophus*. Twelve loci predicted to encode NPRS were deleted singly, and only one, *NPS6*, appeared to be involved in virulence to maize (Oide et al. 2006). Interestingly, *NPS6* is the only gene that has an ortholog in all other ascomycetes examined. To examine the functional conservation of *NPS6*, *NPS6* orthologs were deleted in the rice pathogen *Cochliobolus miyabeanus*, the *Arabidopsis* pathogen *Alternaria brassicicola*, and the wheat/maize/barley pathogen *Fusarium graminearum*.

Lesions caused by *C. heterostrophus nps6* mutants are smaller compared to those of the wild-type. Conidia of the *nps6* mutant germinate normally, form appressoria and penetrate successfully into the host. However, the extent of colonization by this mutant is less than that of the wild-type strain, indicating that deletion of *NPS6* does not cause a defect in pre-penetration growth or penetration efficiency. Lesion sizes upon infection with the *nps6* mutants of *C. miyabeanus* and *A. brassicicola* are also reduced. Similarly, symptom development on wheat spikes inoculated with the *nps6* mutants of *F. graminearum* is much delayed compared with the wild-type strain. Besides their reduced virulence, *nps6* mutants have increased sensitivity to iron depletion and increased sensitivity to oxidative stress. Indeed, *C. heterostrophus*, *Al. brassicicola*, and *F. graminearum NPS6* genes are upregulated under iron-depleted conditions. These observations suggested that the NPRS encoded by *NPS6* are involved in siderophore biosynthesis. HPLC analyses of the culture filtrate and mycelial fractions of *C. heterostrophus* wild-type and *nps6* mutants confirmed this assumption. Further analyses demonstrated that the siderophores produced are coprogens, which are hydroxamates widely distributed among fungal species, including the saprophyte *Neurospora crassa* (Winkelmann 2007). The role of these siderophores in iron nutrition of these fungi *in planta* was supported by the observation that exogenous application of iron enhances the virulence of *nps6* mutants. The application of the extracellular siderophore of *A. brassicicola* restores wild-type virulence to the *nps6* mutant on *Arabidopsis*. Whether these siderophores play a role in virulence through the protection of fungal pathogens against reactive oxygen species generated *in planta* requires further investigation.

The filamentous ascomycete *Magnaporthe grisea* causes rice blast disease, which is very devastating in cultivated rice (Caracuel-Rios and Talbot 2007). Infection occurs through the formation of specialized structures called appressoria that differentiate from the end of fungal germ tubes and generate mechanical force necessary to romper the leaf cuticle and entry to internal tissues. *M. grisea* can also infect roots by development of distinct infection structures. Once inside roots, the fungus can invade the plant vascular system and spread to aerial parts of the plant where it produces disease lesions. *M. grisea* produces extracellular siderophores of the coprogen family to acquire iron from the environment. It also synthesizes the siderophore ferricrocin, a cyclic hexapeptide located intracellularly and considered as an iron storage molecule under iron-replete conditions. In the *M. grisea* genome, Hof et al. (2007) identified a NRPS gene, *SSM1*, clustered with other genes potentially involved in siderophore biosynthesis. Disruption of *SSM1* confirmed that the gene encodes ferricrocin synthase. The ability of the ferricrocin deficient mutants to infect rice leaves was examined in detached leaf assays. On intact leaves, the mutants caused almost no disease symptoms, while on wounded leaves they induced lesions similar to those of the wild-type strain. Using a whole plant infection assay with a susceptible rice cultivar, the authors confirmed that the mutants are impaired in their ability to develop the typical symptom of gray-centered lesion. The mutants showed no defects in term of sporulation, germination of conidia or appressorium formation however, a cell penetration assay revealed that they have reduced ability to penetrate the plant surface. Therefore, ferricrocin appears to play a role in an early plant infection step which is critical for making the appressorium functional.

2.1.8 Ustilago maydis, *a Basidiomycete Model*

The plant-pathogenic basidiomycete *U. maydis* causes smut disease of maize. The main symptoms are tumors which can develop on all green parts of the plant (Bölker 2001). The fungal hyphae proliferate in these tumors and differentiate into diploid spores. Spores are distributed by air and can germinate after landing on plant surfaces. During germination, meiosis occurs and results in the production of nonpathogenic haploid cells. Fusion of compatible haploid cells is required to generate a dikaryotic mycelium which is infectious. The fusion of cells and the development of the pathogenic dikaryon are governed by the a and b mating-type loci. During mating, cells secrete specific pheromones and respond to the presence of cells of the opposite mating type. This process involves coordinated cAMP-dependent kinase A (PKA), as well as mitogen-activated protein kinase (MAPK) signaling. Transcriptome analysis using whole genome microarrays to identify putative targets of the PKA catalytic subunit Adr1 revealed nine genes with putative functions in two high-affinity uptake systems, located in three gene clusters on chromosomes 1, 2, and 4 of the *U. maydis* genome (Eichhorn et al. 2006).

The cluster on chromosome 1 contains the *sid1* and *sid2* genes previously shown to be involved in ferrichrome biosynthesis (Mei et al. 1993). Chromosome 2 harbors a cluster of genes, *fer3* to *fer10* (for Fe-regulated), several of which were predicted to encode the enzymes catalyzing the biosynthesis of ferrichrome A, the second siderophore released by *U. maydis*, as well as the corresponding transporters. The two genes found on chromosome 4, *fer1* and *fer2*, encode a putative ferroxidase and a high-affinity ferric permease, respectively. The *fer2* gene is able to complement the growth defect of an *FTR1* mutant of *Saccharomyces cerevisiae* affected in high-affinity iron uptake due to a lack of functional ferric iron permease. It is plausible that, like in *S. cerevisiae*, the iron permease and oxidase in *U. maydis* are part of a ferrous uptake system in which a plasma membrane reductase reduces ferric ions present in the medium. The resulting ferrous ions are then reoxidized by the multicopper oxidase before transport by the permease. Like the *sid1* and *sid2* genes, the *fer* genes were found to be negatively controlled by the GATA transcription factor Urbs1 in the presence of iron. As these genes are also regulated through a cAMP-dependent pathway, it was postulated that the regulatory activity of Urbs1, which has eight putative PKA phosphorylation sites, could depend on phosphorylation.

When maize plants are infected with a compatible combination of wild-type strains, tumors develop in most infected plants and a high proportion of these plants die. Upon infection with either *fer2* or *fer1* mutants fewer plants develop tumors and these tumors remain small. As a consequence, plant death is only rarely observed. These mutants are not affected in early development on the leaf surface: they form appressoria and penetrate normally. However, proliferation of the fungal cells in the plant is not observed. Interestingly, the *fer2* mutant is more virulent on a wild-type host than on the maize *ys1* mutant, which has a defect in ferric phytosiderophore uptake. Thus, the ferroxidation/permeation system is decisive for iron acquisition during plant colonization by *U. maydis*. As ferrichrome biosynthetic mutants are not affected in virulence, an intriguing question is why phytopathogenic fungi use different strategies for iron acquisition during plant colonization. It is possible that the lifestyle of the fungus during infection, i.e., growing as a biotroph or a necrotroph may determine in which form the iron present in the host plant can be acquired. At which stages of its life cycle *U. maydis* makes use of the ferrichrome-mediated iron uptake system is another remaining question.

The role of the siderophore rhodotulic acid in the phytopathogenesis of *Microbotryum violaceum*, an obligate anther smut fungus infecting species of the Caryophyllacea family, was also investigated by characterizing a mutant deficient in extracellular production of this molecule. Although the growth of this mutant compared to wild-type is altered when starved for iron, the production of rhodotulic acid proved to be unnecessary for fungal growth *in planta* (Birch and Ruddat 2005).

2.2 Iron and Plant Defense

2.2.1 Iron Homeostasis in Wheat Upon Infection by Blumeria graminis

The powdery mildew fungus *Blumeria graminis* is a biotrophic pathogen that requires successful host cell wall penetration, development of a functional haustorium and maintenance of host cell integrity to establish a compatible interaction with its host. Resistance to this pathogen is expressed by prevention of penetration through localized cell wall strengthening by the apposition of new wall material. During *B. graminis* attack, there is production of reactive oxygen species in epidermal and mesophyll cells close to the infection sites. In an expressed sequence tag (EST) library developed from wheat epidermis challenged with the wheat powdery mildew fungus *B. graminis* sp. *tritici*, Liu et al. (2007) noted a high occurrence of Fe-related transcripts, which prompted these authors to investigate changes in iron homeostasis in wheat leaves during the fungal attack.

Using ICPMS (inductively coupled plasma mass spectrometry) to track changes in metal concentrations in the epidermis and mesophyll of wheat leaves infected with *B. graminis*, Liu et al. (2007) found that ferric ions accumulate in the infected epidermis but not in the mesophyll. This accumulation takes place when cell wall appositions are mature. Further analysis indicated that the accumulated iron enhances the production of reactive oxygen species triggered during pathogen attack. Indeed, a pretreatment of the inoculated leaves with the ferric iron chelator desferrioxamine decreased the oxidative burst detected by diaminobenzidine (DAB) staining. The iron-regulated genes *TmNAS1* and *TmFER1*, encoding a nicotianamine synthase and a ferritin isoform, respectively, appeared to be down-regulated during pathogen attack, suggesting cytosolic iron depletion in the infected tissues. In addition, a correlation was established between cytosolic iron depletion and activation of the pathogenesis-related gene *TmPR1b*. Otherwise, using the membrane-permeable fluorescent chelator calcein, the authors showed that a treatment of wheat suspension cells with H_2O_2 promoted active cytosolic Fe efflux. Collectively, these findings led to the proposal of a model of the involvement of iron in wheat leaves during *B. graminis*, where the redistribution of this metal and the oxidative burst triggered after infection contribute to the amplification of a redox signal involved in the defense response.

2.2.2 Involvement of Ferritin in the Response of Potato to Phytophthora infestans

Oomycetes in the genus *Phytophthora* are fungus-like plant pathogens that are devastating for virtually all dicots (Attard et al. 2008). The species *Phytophthora infestans* causes potato late blight and this disease is initiated when wind-blown sporangia release zoospores onto the plant surface. The zoospores encyst to form appressoria, after which hyphae develop and can spread throughout the plant. When these structures penetrate plant cells, they remain enveloped by a lipid membrane derived from the plant plasma membrane. Treatment of potato leaves with desferrioxamine before spraying a suspension of sporangia resulted in a reduction of lesion development and a decrease in the production of reactive oxygen species (Mata et al. 2001). Upon infection with *P. infestans* expression of the *StF1* potato ferritin gene was increased, suggesting that ferritin may be a protective molecule for plant cells. By scavenging the intracellular iron, ferritin can limit the generation of reactive oxygen species. Although increased ferritin gene expression does not suffice to confer resistance to *P. infestans*, a previous study indicated that transgenic tobacco plants ectopically expressing an alfalfa ferritin gene has increased tolerance to viral (tobacco necrosis virus) and fungal (*Alternaria alternata*, *Botrytis cinerea*) infections (Deak et al. 1999). These reports support the idea that ferritin can be part of host defense responses triggered during infection.

2.2.3 Iron Homeostasis and Resistance of Arabidopsis to **D. dadantii**

As efficient iron uptake/storage mechanisms are of critical importance to the survival of *D. dadantii* during infection, the question of whether this pathogen could induce iron-withholding reactions in the plant has been addressed. A first observation made by Neema et al. (1993) indicated that iron incorporated into plant ferritins drastically decreased in soybean suspension cells challenged with *D. dadantii*. This effect was also observed during treatment of the cells with chrysobactin. The possibility of a competition for iron between the pathogen and the host was also illustrated by the observation that accumulation of polyphenols in plant tissues inhibits growth of mutants of *D. dadantii* affected in their sidero-phore-mediated iron transport pathway (Mila et al. 1996). Polyphenols have iron-chelating properties and could play the role fulfilled by iron-binding proteins, such as transferrin in animal immunity.

In order to identify plant genes that are regulated in response to infection by *D. dadantii*, Dellagi et al. (2005) differentially screened cDNA libraries from *Arabidopsis* and found that the gene encoding the ferritin AtFer1 is upregulated in infected plants. These authors established that accumulation of *AtFer1* transcripts and production of ferritins during infection is a defense reaction against prolifer-ation of the pathogen. The siderophore chrysobactin, as well as desferrioxamine are elicitors of this response. As only the iron-free siderophores induce this reaction, it was suggested that these iron sequestering molecules could cause severe iron depletion in *Arabidopsis* leaf tissues, resulting in the redistribution of intracellular iron stores and/or the activation of iron acquisition systems of the cell. Intracellular redistribution could involve remobilization of vacuolar iron by the specific metal transporters Nramp3 and Nramp4 (Fig. 2.2). Indeed, among the six *NRAMP* genes present in *Arabidopsis*, *AtNramp3* was found to be strongly upregulated in response to several biotic stresses.

Because there is a functional redundancy between AtNRAMP3 and AtNRAMP4 in seed germination and the encoded proteins share 50 % sequence identity with the mouse NRAMP1 metal ion transporter involved in innate immunity, the role of these genes in resistance to *D. dadantii* was investigated further (Segond et al. 2009). *AtNRAMP3* is upregulated in leaves challenged with *D. dadantii*, while *AtNRAMP4* expression does not change. Using simple and double *nramp3* and *nramp4* mutants, as well as lines ectopically expressing either of these genes, Segond et al. (2009) showed that *AtNRAMP3*, and to a lesser extent *AtNRAMP4*, are involved in the resistance of *Arabidopsis* against this bacterium. The susceptibility of the *nramp3 nramp4* double mutant was associated with a reduced accumulation of reactive oxygen species and AtFER1, which are effective defense components against *D. dadantii*. By promoting an efflux of iron and possibly other metals from the vacuole to the cytosol, the activity of Nramp3 and Nramp4 proteins may contribute to exacerbation of the oxidative stress generated during infection and be responsible for basal resistance to this pathogen. Both increased oxidative stress and efflux of

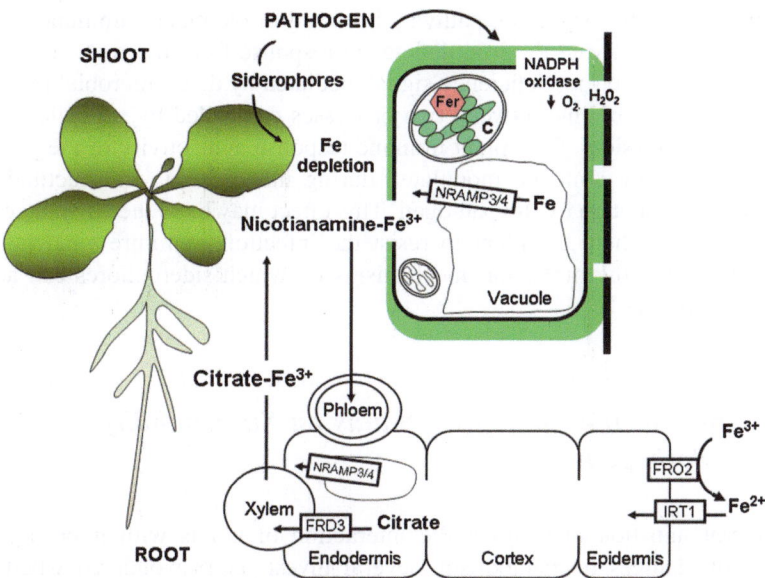

Fig. 2.2 Schematic representation of plant iron acquisition and changes in iron trafficking during pathogenesis. This model is based on the reactions triggered by *D. dadantii* during infection of *Arabidopsis*. Plants acquire iron from the soil. In dicots, Fe^{3+} is reduced to Fe^{2+} by the FRO2 ferric chelate reductase, and then transported through the plasma membrane by the iron-regulated transporter IRT1. Inside the plant, iron is transported essentially as ferric complexes of citrate in the xylem and of nicotianamine in the phloem. Storage and buffering occur in the apoplast and the organelles including vacuoles and plastids (C) that contain ferritins (Fer). Bacterial invasion triggers iron depletion in leaves, leading to a mobilization of vacuolar iron mediated by transporters AtNramp3 and AtNramp4. The reactive iron released in the cytosol contributes to amplify the production of reactive oxygen species and to induce ferritin synthesis in the chloroplast depriving the bacteria of iron. Infection also results in iron mobilization in the roots, from both the vacuole and the soil

iron could be at the origin of ferritin up-regulation in infected leaves. In addition, roots from *D. dadantii*-challenged plants accumulate transcripts of *AtNRAMP3* as well as the root iron deficiency markers *IRT1* and *FRO2*. This finding suggests the existence of a shoot to root signal activated by pathogen infection. Whether the redistribution of iron in the infected leaf and uptake of iron by the roots are physiologically linked is an appealing question. Collectively, these data indicate that the functions of NRAMP proteins in innate immunity (Neves et al. 2011) have been conserved between animals and plants.

Are the siderophores produced by the bacterium the main elicitors of these plant reactions? To approach this question, Dellagi et al. (2009) analyzed the effect of diverse types of siderophores including chrysobactin and desferrioxamine in Arabidopsis plants following leaf infiltration. It was found that these siderophores, only when iron-free, could activate the salicylic acid-dependent reactions belonging to a major signaling pathway involved in the plant's immune network. They also could activate the iron deficiency response from the root, showing the existence of a leaf to root deficiency signal mediated by these microbial molecules, and revealing a new link between two processes controlled by salicylic acid and iron. Thus, expression of the plant immune response and activity of the plant iron acquisition system could be modulated during infection via the fluctuations of siderophore production by the pathogen. This effect may be to the advantage of the pathogen or may help the plant to resist the infection. A future challenge is to better understand the molecular mechanisms by which siderophores can activate this dual response.

2.2.4 Effect of the Plant Iron Status on Susceptibility/Resistance to Pathogens

As mineral nutrition may affect the interaction of plants with microorganisms because of changes in metabolism, several investigators wondered whether the iron status of the host could exert a significant effect on the disease evolution. Indeed, iron deficiency can compromise the activity of metalloenzymes important for the host immune response. This question was addressed in several studies in which plants grown under iron deficiency conditions were challenged with fungal pathogens. In particular, solanaceous crop plants were infected by species of *Fusarium* and *Verticillium* which are ascomycetes responsible for wilt diseases. These fungi usually enter the plant through young roots and then grow into the water conducting vessels of the roots and stem. As the vessels are plugged and collapse, the water supply to the leaves is blocked. Anderson and Guerra (1985) observed increased lesion size in beans infected by *F. solani* by reducing iron in the nutrient solution of plants grown under hydroponic conditions. In infection by *V. dalhliae* of several crops, including peanuts, egg-plants, tomato, and potato, iron deficiency resulted in increased sensitivity to symptom development. Supplementation of plants with iron significantly alleviated disease symptoms (Barash 1988), however in peanuts, this effect was not correlated with production of phytoalexins, a typical antimicrobial plant reaction. Studies recently conducted on *Arabidopsis* with two aerial pathogens *D. dadantii* and the ascomycete *Botrytis cinerea*, showed a different result (Kieu et al. 2012). Iron-starved plants displayed reduced susceptibility to infection and iron supplementation restored the symptom severity. In the case of *D. dadantii*, further examination revealed that iron deficiency causes a reduction in bacterial fitness and expression of virulence genes as well as an exacerbation of the salicylic acid-mediated defense pathway. But disease reduction did not correlate with the involvement of defenses known to be effective against the bacterium. Thus, the plant iron status can influence host-pathogen relationships in different ways by affecting the pathogen's virulence as well as the host's defense.

2.3 Conclusion

Like other members of the microbial world interacting with animals and humans, plant pathogenic microorganisms have evolved a diversity of systems allowing them to capture iron from various environments in response to their metabolic needs. Indeed, experimental investigations have permitted to underscore the importance of siderophores and corresponding transport machineries in iron nutrition of phytopathogenic species. Moreover, knowledge of complete sequences of bacterial and fungal genomes has greatly contributed to enlarge our vision of how plant pathogens can manage their iron homeostasis. These microbes produce siderophores belonging to the three broad classes, hydroxamate, catecholate, and α-hydroxycarboxylate as chelating groups. Bacterial phytopathogens are also gifted with specific transport systems for ferrous iron, ferric citrate, and other sources of ferric iron, heme and exogenous microbial siderophores (Table 2.1). Reductive iron uptake mechanisms are likely to exist in plant pathogenic fungi, but the possibilities of transporting heme and ferric ligands other than siderophores have not been reported. The plurality of these systems illustrates the importance of iron acquisition to the fitness of the bacteria associated to plants. Outside the plant, the natural habitats of pathogens are the rhizosphere and soils where many microbial populations interact with each other or with the plant. Very likely, the possibility to express well-suited mechanisms for iron acquisition with the capacity to utilize exogenous siderophores is essential for their survival.

Although the potentiality to produce siderophores seems to be a common trait to most phytopathogens, the requirement of a functional siderophore-mediated iron transport route for pathogenicity is not absolute. As siderophores are powerful iron-chelating compounds and can compete out with known plant iron transporters including organic acids and nicotianamine, or with iron containing proteins, this indicates that in certain pathological situations, the availability of this metal is sufficient to enable the pathogen to thrive *in planta*. Indeed, depending on the infected organ or tissue and the strategy of microbial attack, it might be beneficial for the pathogen to use alternative iron transport routes or a heme acquisition system. A preference for the Feo ferrous iron acquisition system was demonstrated in the case of *X. oryzae* infection and a ferric iron permease is required for the virulence of *U. maydis*. Production of a siderophore dependent iron transport system requires the involvement of NRPS and a specific TonB-dependent transporter and the whole process is metabolically energy costing. Secretion of virulence factors is likely to be expensive in energy as well and co-expression of both functions could be meta-bolically incompatible. The fact that expression of the *P. syringae* DC3000 type III secretion system in culture requires sufficient iron levels supports this assumption. In the same way, several genes encoding the *D. dadantii* pectate-lyase involved in the virulence are repressed when the host is iron-depleted. The iron sensory protein Fur was shown to act as a global regulator of phytopathogenicity in several pathosystems and the discovery of new virulence regulatory networks responding to iron levels highlights the importance of a tight control of the bacterial intracellular iron level in relation to metabolic activities during infection. It is well known that during the infectious process, phytopathogenic bacteria encounter an oxidative environment. Reactive oxygen species are generated by host plants as a defense mechanism against microbial invasion. This plant defense response consists of the production of superoxide, hydrogen peroxide, and nitric oxide, which function either directly in the establishment of defense mechanisms or indirectly via synergistic interactions with other signaling molecules, such as salicylic acid (reviewed in Bolwell and Daudi 2009). Under these conditions, a tight control of the iron concentration is essential for the invading bacteria to avoid exacerbation of this oxidative stress through Fenton's reaction, which generates the highly toxic and reactive hydroxyl radical OH·. Interestingly, siderophores can interfere with this reaction by sequestering ferric ions and thus influencing the ferric/ferrous iron ratio; for instance, desferrioxamine produced by *E. amylovora* was proposed to function as such.

An obstacle in the comprehension of iron competition between bacteria and plants is the lack of information on the siderophores produced by phytopathogenic bacteria. Very few siderophores have been chemically and structurally characterized in plant pathogenic species (Table 2.2). Depending on their iron-chelating functional groups, siderophore molecules possess different affinity for the ferric ion and exhibit different chelating capacity and stability under pH variations. These parameters must be evaluated in order to determine the functional role that a given siderophore can play during the infectious cycle. Several phytopathogenic bacteria encode more than one siderophore iron uptake system that can be distinct in very close strains or pathovars. An illustration of this complexity is illustrated by the strain EC16 of *D. dadantii* which produces dichrysobactin and linear/cyclic trichrysobactin in addition to the monomeric siderophore chrysobactin (Sandy and Butler 2011). Synthesis of different siderophores may help bacteria to cope with fluctuations of the iron status encountered within plant tissues. Indeed, different siderophore dependent iron transport routes can be differentially expressed and play distinct roles according to environmental conditions or pathological situations. These aspects are important to further consider the role of siderophores in pathogenicity.

The role of iron in microbial plant pathogenesis is an emerging topic with exciting recent developments. There is now evidence that a competition for iron between the host and the microorganism can take place. In particular, it is noteworthy that during infection by *D. dadantii*, Arabidopsis plants develop an iron-withholding response that changes iron distribution and trafficking. This effect is likely to be caused by the *D. dadantii* siderophores that can interfere with plant defense responses and exacerbate iron mobilization from the root. The fact that the plant nutritional iron status also affects the development of diseases, by acting on both the pathogen's virulence and the host's defense is demonstrative. Elucidating mechanisms involved in exchanging and withholding iron during plant–microbe interactions should help to develop integrative strategies for controlling plant diseases. Such investigations are of primary importance in an agronomical context, where reductions in mineral fertilization and pesticide crop treatment are becoming necessary. They should help to the development of crop genotypes with

Table 2.2   Siderophores produced by phytopathogenic bacteria, with their role in virulence

Common bidentate coordination sites in microbial siderophores

| Catechol | Hydroxamic acid | Hydroxy carboxylic acid |

Species/Strain	Typical disease	Characterized siderophore	Role in pathogenicity (reference)
P. syringae pv. *syringae* strain B301D	Fruit necrotic spots	Pyoverdin	Not required for cherry fruit disease (Cody and Gross 1987)
P. syringae pv. *tomato* strain DC3000	Bacterial speck of tomato	Yersiniabactin Pyoverdin	Not required for tomato and Arabidopsis diseases (Jones et al. 2007) (Jones and Wildermuth 2011)
P. syringae pv. *tabaci* strain 6605	Wildfire disease on host tobacco plants	Pyoverdin	Necessary for tobacco leaves infection (Taguchi et al. 2010)
P. syringae pv. *phaseolicola* strain 1448a	Bean halo blight	Pyoverdin and Achromobactin	Unnecessary for bean pods infection (Owen and Ackerley 2011)
A. tumefaciens strain B6	Crown gall tumors on dicot plants	Agrobactin	Not required for carrot or sunflower disease (Leong and Neilands 1981)
R. solanacearum strain AW1	Bacterial wilt on many plants	Staphyloferrin B	Unnecessary for tomato plants infection (Bhatt and Denny 2004)
E. chrysanthemi (*D. dadantii*) strain 3937	Soft rotting on many plants	Chysobactin and achromobactin	Required for virulence on host plants (Enard et al. 1988; Neema et al. 1993; Franza et al. 2005)
E. carotovora strain W3C105	Potato stem rot	Chrysobactin and aerobactin	No role in aerial stem rot (Ishimaru and Loper 1992)
E. amylovora strain 1430	Fire bight on *Pomoideae*	Deferrioxamines	Necessary for virulence apple flowers (Dellagi et al. 1998)

efficient mineral use and uptake properties without an increase in susceptibility to plant pathogens and/or decrease in yield or crop quality. In addition, the use of naturally occurring microbial antagonists to suppress plant diseases offers an alternative to classical methods of plant protection and in this regard, it is worth considering that production of high-affinity iron transport systems by such antagonists can be a determining factor in their competitiveness.

Furthermore, iron deficiency represents a major nutritional disorder and is estimated to affect a great part of the world's population. A commonly used strategy to increase dietary intake of iron is to develop iron biofortified plants (Murgia et al. 2012). Potential changes in susceptibility of plants to microbial diseases must be considered, when biotechnological approaches aimed at enhancing the bioavailability of iron in the tissues are developed.

Acknowledgments Research from the authors' laboratory was supported by grants from the INRA and the Université Pierre et Marie Curie (Paris)

References

Attard A, Gourgues M, Galiana E, Panabières F, Ponchet M, Keller H (2008) Strategies of attack and defense in plant-oomycete interactions, accentuated for *Phytophtora parasitica* Dastur (Syn. P. Nicotianae Breda de Haan). J Plant Physiol 165:83–94

Barash I, Zion R, Krikun J, Nachmias A (1988) Effect of iron status on Verticillium wilt disease and on in vitro production of siderophores by *Verticillium dahliae*. J Plant Nutr 11:893–905

Barnes HH, Ishimaru CA (1999) Purification of catechol siderophores by boronate affinity chromatography: identification of chrysobactin from *Erwinia carotovora* subsp. *carotovora*. Biometals 12:83–87

Berti AD, Thomas MG (2009) Analysis of achromobactin biosynthesis by *Pseudomonas syringae* pv. *syringae* B728a. J Bacteriol 191:4594–4604

Bhatt G, Denny TP (2004) *Ralstonia solanacearum* iron scavenging by the siderophore staphyloferrin B is controlled by PhcA, the global virulence regulator. J Bacteriol 186:7896–7904

Birch LE, Ruddat M (2005) Siderophore accumulation and phytopathogenicity in *Microbotryum violaceum*. Fungal Gent Biol 42:579–589

Bölker M (2001) *Ustilago maydis*, a valuable model system for the study of fungal dimorphism and virulence. Microbiology 147:1395–1401

Bolwell GP, Daudi A (2009) Reactive Oxygen species in plant signaling. In: del Rio LA, Puppo A (eds) Signaling and Communication in Plants. Springer, Berlin, pp 113–133

Boughammoura A, Expert D, Franza T (2012) Role of the *Dickeya dadantii* Dps protein. Biometals 25:423–433

Boughammoura A, Matzanke BF, Böttger L, Reverchon S, Lesuisse E, Expert D, Franza T (2008) Differential role of ferritins in iron metabolism and virulence of the plant pathogenic bacterium *Erwinia chrysanthemi* 3937. J Bacteriol 190:1518–1530

Briat JF, Duc C, Ravet K, Gaymard F (2010) Ferritins and iron storage in plants. Biochim Biophys Acta 1800:806–814

Bultreys A, Gheysen I, de Hoffmann E (2006) Yersiniabactin production by *Pseudomonas syringae* and *Escherichia coli* and description of a second yersiniabactin locus evolutionary group. App Environ Microbiol 92:3814–3825

Butcher BG, Bronstein PA, Myers CR, Stodghill PV, Bolton JJ, Markel E, Filiatrault MJ, Swingle B, Gaballa A, Helmann JD, Schneider DJ, Cartinhour S (2011) Characterization of the Fur regulon in *Pseudomonas syringae* pv. *tomato* DC3000. J Bacteriol 193:4598–4611

Caracuel-Rios Z, Talbot NJ (2007) Cellular differentiation and host invasion by the rice blast fungus *Magnaporthe grisea*. Curr Opin Microbiol 10:339–345

Cha JY, Lee JS, Oh JI, Choi JW, Baik HS (2008) Functional analysis of the role of Fur in the virulence of *Pseudomonas syringae* pv. *tabaci* 11528: Fur controls expression of genes involved in quorum-sensing. Biochem Biophys Res Commun 366:281–287

Chatterjee S, Sonti RV (2002) *rpfF* mutants of *Xanthomonas oryzae* pv. *oryzae* are deficient for virulence and growth under low iron conditions. Mol Plant Microbe Interact 15:463–471

Chatterjee S, Almeida RPP, Lindow S (2008) Living in two worlds: the plant and insect lifestyles of *Xylella fastidiosa*. Annu Rev Phytopathol 46:243–271

Chiancone E, Ceci P (2010) The multifaceted capacity of Dps proteins to combat bacterial stress conditions: detoxification of iron and hydrogen peroxide and DNA binding. Biochem Biophys Acta 1800:798–805

Cody YS, Gross DC (1987) Characterization of pyoverdin$_{pss}$, the fluorescent siderophore produced by *Pseudomonas syringae* pv. *syringae*. App Environ Microbiol 53:928–934

Colburn-Clifford JM, Scherf JM, Allen C (2010) Ralstonia solanacearum Dps contributes to oxidative stress tolerance and to colonization of and virulence on tomato plants. App Environ Microbiol 76:7392–7399

Crosa JH, Mey AR, Payne SM (2004) Iron transport in bacteria: molecular genetics, biochemistry, microbial pathogenesis and ecology. American Society of Microbiology (ASM) Press Book, Washington

Curie C, Cassin G, Couch D, Divol F, Higuchi K, Le Jean M, Misson J, Schikora A, Czernic P, Mari S (2009) Metal movement within the plant: contribution of nicotianamine and yellow stripe 1-like transporters. Ann Bot 103:1–11

Czajkowski R, Pérombelon MCM, van Veen JA, van der Wolf JM (2011) Control of blackleg and tuber soft rot of potato caused by *Pectobacterium* and *Dickeya* species: a review. Plant Pathol 60:999–1013

Deak M, Horvath GV, Davletova S, Torok K, Sass L, Vass I, Barna B, Kirali Z, Dudits D (1999) Plants ectopically expressing the iron-binding protein, ferritin, are tolerantto oxidative damage and pathogens. Nat Biotechnol 17:192–196

Degrave A, Fagard M, Perino C, Brisset MN, Gaubert S, Laroche S, Patrit O, Barny MA (2008) *Erwinia amylovora* type three-secreted proteins trigger cell death and defense responses in *Arabidopsis thaliana*. Mol Plant Microbe Interact 21:1076–1086

Dellagi A, Brisset MN, Paulin J-P, Expert D (1998) Dual role of desferrioxamine in *Erwinia amylovora* pathogenicity. Mol Plant Microbe Interact 8:734–742

Dellagi A, Reis D, Vian B, Expert D (1999) Expression of the ferrioxamine receptor gene of *Erwinia amylovora* CFBP 1430 during pathogenesis. Mol Plant Microbe Interact 12:463–466

Dellagi A, Rigault M, Segond D, Roux C, Kraepiel Y, Cellier F, Briat JF, Gaymard F, Expert D (2005) Siderophore-mediated upregulation of Arabidopsis ferritin expression in response to *Erwinia chrysanthemi* infection. Plant J 43:262–272

Dellagi A, Segond D, Rigault M, Fagard M, Simon C, Saindrenan P, Expert D (2009) Microbial siderophores exert a subtle role on Arabidopsis during infection by manipulating the immune response and the iron status. Plant Physiol 150:1687–1696

Eatsgate JA (2000) *Erwinia amylovora*, the molecular basis of fireblight disease. Mol Plant Pathol 1:325–332

Eichhorn H, Lessing F, Winterberg B, Schirawski J, Kämper J, Müller P, Kahmann R (2006) A ferroxidation/permeation iron uptake system is required for virulence in *Ustilago maydis*. Plant Cell 18:3332–3345

Enard C, Diolez A, Expert D (1988) Systemic virulence of *Erwinia chrysanthemi* 3937 requires a functional iron assimilation system. J Bacteriol 170:2419–2426

Etchegaray A, Silva-Stenico ME, Moon DH, Tsai SM (2004) In silico analysis of nonribosomal peptide synthetases of *Xanthomonas axonopodis* pv. *citri*:identification of putative siderophore and lipopeptide biosynthetic genes. Microbiol Res 159:425–437

Expert D, Boughammoura A, Franza T (2008) Siderophore controlled iron assimilation in the enterobacterium *Erwinia chrysanthemi*: evidence for the involvement of bacterioferritin and the Suf iron-sulfur cluster assembly machinery. J Biol Chem 283:36564–36572

Expert D, Rauscher L, Franza T (2004) *Erwinia*, a plant pathogen. In: Crosa JH, Mey AR, Payne SM (eds) Iron transport in bacteria: molecular genetics, biochemistry, microbial pathogenesis and ecology. American Society of Microbiology (ASM) Press Book, Washington, pp 402–412

Expert D, Toussaint A (1985) Bacteriocin-resistant mutants of *Erwinia chrysanthemi*: possible involvement of iron acquisition in phytopathogenicity. J Bacteriol 163:221–227

Fagard M, Dellagi A, Roux C, Périno C, Rigault M, Boucher V, Shevchik V, Expert D (2007) *Arabidopsis thaliana* expresses multiple lines of defense to counter-attack *Erwinia chrysanthemi*. Mol Plant Microbe Interact 20:794–805

Feil H, Feil WS, Chain P, Larimer F, DiBartolo G, Copeland A, Lykidis A, Trong S, Nolan M, Goltsman E, Thiel J, Malfatti S, Loper JE, Lapidus A, Detter JC, Land M, Richardson PM, Kyrpides NC, Ivanova N, Lindow SE (2005) Comparison of the complete genome sequences of *Pseudomonas syringae* pv. *syringae* B728a and pv. *tomato* DC3000. Proc Natl Acad Sci 102:11064–11069

Feistner GJ, Stahl DC, Gabrik AH (1993) Proferrioxamine siderophores of *Erwinia amylovora*. A capillary liquid chromatographic/electrospray tandem mass spectrometry study. Org Mass Spectrom 28:163–175

Franza T, Expert D (1991) The virulence-associated chysobactin iron uptake system of *Erwinia chrysanthemi* 3937 involves an operon encoding transport and biosynthetic functions. J Bacteriol 173:6874–6881

Franza T, Expert D (2010) Iron uptake in soft rot Erwinia. In: Cornelis P, Andrews SC (eds) Iron uptake and homeostasis in microorganisms. Caister Academic Press, Norwich, pp 101–115

Franza T, Mahé B, Expert D (2005) *Erwinia chrysanthemi* requires a second iron transport route dependent of the siderophore achromobactin for extracellular growth and plant infection. Mol Microbiol 55:261–275

Franza T, Michaud-Soret I, Piquerel P, Expert D (2002) Coupling of iron assimilation and pectinolysis in *Erwinia chrysanthemi* 3937. Mol Plant Microbe Interact 15:1181–1191

Franza T, Sauvage C, Expert D (1999) Iron regulation and pathogenicity in *Erwinia chrysanthemi* strain 3937: Role of the Fur repressor protein. Mol Plant-Microbe Interact 12:119–129

Genin S (2010) Molecular traits controlling host range and adaptation to plants in *Ralstonia solanacearum*. New Phytol 187:920–928

Greenwald JW, Greenwald CJ, Philmus BJ, Begley TP, Gross DC. (2012) RNA-seq analysis reveals that an ECF σ Factor, AcsS, regulates achromobactin biosynthesis in *Pseudomonas syringae* pv. *syringae* B728a. PLoS One 7(4):e34804 (Epub 2012 Apr 18)

Gross DC (1985) Regulation of syringomycin synthesis in *Pseudomonas syringae* pv. *syringae* and defined conditions for its production. J Appl Bacteriol 58:167–174

Guerra D, Andreson AJ (1985) The effect of iron and boron amendments on infection of bean by *Fusarium solani*. Phytopathology 75:989–991

Hernandez-Morales A, De la Torre-Zavala S, Ibarra-Laclette E, Hernandez-Flores JL, Jofre-Garfias AE, Martinez-Antonio A, Alvarez-Morales A (2009) Transcriptional profile of *Pseudomonas syringae* pv. *phaseolicola* NPS3121 in response to tissue extracts from a susceptible *Phaseolus vulgaris* L. cultivar. BMC Microbiol 9:257

Hibbing ME, Fuqua C (2011) Antiparallel and interlinked control of cellular iron levels by the Irr and RirA regulators of *Agrobacterium tumefaciens*. J Bacteriol 193:3461–3472

Hof C, Eisfeld K, Welzel K, Antelo L, Foster AJ, Anke H (2007) Ferricrocin synthesis in *Magnaporthe grisea* and its role in pathogenicity in rice. Mol Plant Pathol 8:163–172

Jittawuttipoka T, Sallabhan R, Vattanaviboon P, Fuangthong M, Mongkolsuk S (2010) Mutations of ferric uptake regulator (fur) impair iron homeostasis, growth, oxidative stress survival, and virulence of *Xanthomonas campestris* pv. *campestris*. Arch Microbiol 192:331–339

Jones AM, Lindow SE, Wildermuth MC (2007) Salicylic acid, yersiniabactin, and pyoverdin production by the model phytopathogen *Pseudomonas syringae* pv. *tomato* DC3000: synthesis, regulation, and impact on tomato and Arabidopsis host plants. J Bacteriol 189:6773–6786

Jones AM, Wildermuth MC (2011) The phytopathogen *Pseudomonas syringae* pv. *tomato* DC3000 has three high-affinity iron-scavenging systems functional under iron limitation conditions but dispensable for pathogenesis. J Bacteriol 193:2767–2775

Kachadourian R, Dellagi A, Laurent J, Bricard L, Kunesch G, Expert D (1996) Desferrioxaminedependent- iron transport in Erwinia amylovora CFBP1430: cloning of the gene encoding the ferrioxamine receptor FoxR. Biometals 9:143–150

Kieu NP, Aznar A, Segond D, Rigault M, Simond-Cote E, Kunz C, Soulie M-C, Expert D, Dellagi A (2012) Iron deficiency affects plant defense responses and confers resistance to *Dickeya dadantii* and *Botrytis cinerea*. Mol Plant Pathol (in press)

Kim BJ, Park JH, Bronstein PA, Schneider DJ, Catinhour SW, Schuler ML (2009) Effect of iron concentration on the growth rate of *Pseudomonas syringae* and the expression of virulence factors in hrp-inducing minimal medium. App Environ Microbiol 75:2720–2726

Kim BJ, Schneider DJ, Catinhour SW, Shuler ML (2010) Complex responses to culture conditions in *Pseudomonas syringae* pv. *tomato* DC3000 continuous cultures: the role of iron in cell growth and virulence factor induction. Biotechnol Bioeng 105:955–964

Kitphati W, Ngok-Ngam P, Suwanmaneerat S, Sukchawalit R, Mongkolsuk S (2007) *Agrobacterium tumefaciens fur* has important physiological roles in iron and manganese homeostasis, the oxidative stress response, and full virulence. Appl Environ Microbiol 73:4760–4768

Kreutzer MF, Kage H, Gebhardt P, Wackler BW, Saluz HP, Hoffmeister D, Nett M (2011) Biosynthesis of a complex yersiniabactin-like natural product via the *mic* locus in phytopathogen *Ralstonia solanacearum*. App Environ Microbiol 77:6117–6124

Lanquar V, Lelievre F, Bolte S, Hames C, Alcon C, Neumann D, Vansuyt G, Curie C, Schröder A, Krämer U, Barbier-Brygoo H, Thomine S (2005) Mobilization of vacuolar iron by AtNRAMP3 and AtNRAMP4 is essential for seed germination on low iron. EMBO J 24:4041–4051

Lee BN, Kroken S, Chou DYT, Robbertse B, Yoder OC, Turgeon BG (2005) Functional analysis of all non ribosomal peptide synthetases in *Cochliobolus heterostrophus* reveals a factor NPS6, involved in virulence and resistance to oxidative stress. Eucaryotic Cell 4:545–555

Leong SA, Neilands JB (1981) Relationship of siderophore-mediated iron assimilation to virulence in crown gall disease. J Bacteriol 147:482–491

Liu G, Greenshields DL, Sammynaiken R, Hirji RN, Selvaraj G, Wei Y (2007) Targeted alterations in iron homeostasis underlie plant defense responses. J Cell Sci 120:596–605

Loprasert S, Sallabhan R, Atichartpongkul S, Mongkolsuk S (1999) Characterization of a ferric uptake regulator (*fur*) gene from *Xanthomonas campestris* pv. *phaseoli* with unusual primary structure, genome organization, and expression patterns. Gene 239:251–258

Mahé B, Masclaux C, Rauscher L, Enard C, Expert D (1995) Differential expression of two siderophore dependent iron acquisition pathways in *Erwinia chrysanthemi* 3937: characterization of a novel ferrrisiderophore permease of the ABC transporter family. Mol Microbiol 18:33–43

Markel E, Maciak C, Butcher BG, Myers CR, Stodghill P, Bao Z, Cartinhour S, Swingle B (2011) An extracytoplasmic function sigma factor-mediated cell surface signalling system in *Pseudomonas syringae* pv. *tomato* DC3000 regulates gene expression in response to heterologous siderophores. J Bacteriol 193:5775–5783

Mata CG, Lamattina L, Cassia RO (2001) Involvement of iron and ferritin in the potato: *Phytophtora infestans* interaction. Eur J Plant Pathol 107:557–562

Mei B, Budde AD, Leong SA (1993) Sid1, a gene initiating siderophore biosynthesis in Ustilago maydis: molecular characterization, regulation by iron, and role in phytopathogenicity. Proc Natl Acad Sci USA 90:903–907

Meyer JM, Gruffaz C, Tulkki T, Izard D (2008) Taxonomic heterogeneity, as shown by siderotyping, of strains primarily identified as *Pseudomonas putida*. Int J Syst Evol Microbiol 57:2543–2556

Mila I, Scalbert A, Expert D (1996) Iron withholding by plant polyphenols and resistance to pathogens and rots. Phytochemistry 42:1551–1555

Morrissey J, Guerinot ML (2009) Iron uptake and transport in plants: the good, the bad, and the ionome. Chem Rev 109:4553–4567

Münzinger M, Budzikiewicz H, Expert D, Enard C, Meyer JM (2000) Achromobactin, a new citrate siderophore of *Erwinia chrysanthemi*. Z Naturforsch (C) 55:328–332

Murdoch L, Corbel JC, Reis D, Bertheau Y, Vian B (1999) Differential cell wall degradation by *Erwinia chrysanthemi* in petiole of *Saintpaulia ionantha*. Protoplasma 210:59–74

Murgia I, Arosio P, Tarantino D, Suoave C (2012) Biofortification for combating 'hidden hunger' for iron. Trends Plant Sci 17:47–55

Nachin L, El Hassouni M, Loiseau L, Expert D, Barras F (2001) SoxR-dependent response to oxidative stress and virulence of *Erwinia chrysanthemi*: the key role of SufC, an orphan ABC ATPase. Mol Microbiol 39:960–972

Neema C, Laulhere JP, Expert D (1993) Iron deficiency induced by chrysobactin in Saintpaulia leaves inoculated with *Erwinia chrysanthemi*. Plant Physiol 102:967–973

Neves JV, Wilson JM, Kuhl H, Reinhardt R, Castro LF, Rodrigues PN (2011) Natural history of SLC11 genes in vertebrates: tales from the fish world. BMC Evol Biol 11:106

Ngok-Ngam P, Ruangkiattikul N, Mahavihakanont A, Virgem SS, Sukchawalit R, Mongkolsuk S (2009) Roles of *Agrobacterium tumefaciens* RirA in iron regulation, oxidative stress response, and virulence. J Bacteriol 191:2083–2090

Nouet C, Motte P, Hanikenne M (2011) Chloroplastic and mitochondrial metal homeostasis. Trends Plant Sci 16:395–404

O'Brian MR, Fabiano E (2010) Mechanisms and regulation of iron homeostasis in the Rhizobia. In: Cornelis P, Andrews SC (eds) Iron uptake and homeostasis in microorganisms. Caister Academic Press, Norfolk, pp 37–63

Oide S, Moeder W, Krasnoff S, Gibson D, Haas H, Yoshioka K, Turgeon BG (2006) NPS6, encoding a nonribosomal peptide synthetase involved in siderophore-mediated iron metabolism, is a conserved virulence determinant of plant pathogenic ascomycetes. Plant Cell 10:2836–2853

Okinaka Y, Yang CH, Perna NT, Keen NT (2002) Microarray profiling of *Erwinia chrysanthemi* 3937 genes that are regulated during plant infection. Mol Plant Microbe Interact 15:619–629

Ong SA, Peterson T, Neilands JB (1979) Agrobactin, a siderophore from *Agrobacterium tumefaciens*. J Biol Chem 254:1860–1865

Owen JG, Ackerley DF (2011) Characterization of pyoverdine and achromobactin in *Pseudomonas syringae* pv. *phaseolicola* 1448a. BMC Microbiol 11:218

Pandey A, Sonti RV (2010) Role of the FeoB protein and siderophore in promoting virulence of *Xanthomonas oryzae* pv. *oryzae* on rice. J Bacteriol 192:3187–3203

Perry RD, Fetherston JD (2011) Yersiniabactin iron uptake: mechanisms and role in *Yersinia pestis* pathogenesis. Microbes Infect 13:808–817

Persmark M, Expert D, Neilands JB (1989) Isolation, characterization and synthesis of chrysobactin, a compoundwith a siderophore activity from *Erwinia chrysanthemi*. J Biol Chem 264:3187–3193

Pitzchke A, Hirt H (2010) New insights into an old story: *Agrobacterium*-induced tumor formation in plants by plant transformation. EMBO J 29:1021–1032

Rauscher L, Expert D, Matzanke BF, Trautwein AX (2002) Chrysobactin-dependent iron acquisition in Erwinia chrysanthemi: functional study of an homologue of the Escherichia coli ferric enterobactin esterase. J Biol Chem 277:2385–2395

Rondon MR, Ballering KS, Thomas MG (2004) Identification and analysis of a siderophore biosynthetic gene cluster from *Agrobacterium tumefaciens* C58. Microbiology 150:3857–3866

Ryan RP, Vorhölter FJ, Potnis N, Jones JB, Van Sluys MA, Bogdanove AJ, Dow JM (2011) Pathogenomics of *Xanthomonas*: understanding bacterium-plant interactions. Nat Rev Microbiol 9:344–355

Sandy M, Butler A (2011) Chrysobactin siderophores produced by *Dickeya chrysanthemi* EC16. J Nat Prod 74:1207–1212

Santos R, Franza T, Laporte ML, Sauvage C, Touati D, Expert D (2001) Essential role of superoxide dismutase on the pathogenicity of *Erwinia chrysanthemi* strain 3937. Mol Plant-Microbe Interact 14:758–757

Schmelz S, Botting CH, Song L, Kadi N, Challis GL, Naismith JH (2011) Structural basis for acyl acceptor specificity in the achromobactin biosynthetic enzyme AcsD. J Mol Biol 412:495–504

Schmelz S, Kadi N, McMahon SA, Song L, Oves-Costales D, Oke M, Liu H, Johnson KA, Carter LG, Botting CH, White MF, Challis GL, Naismith JH (2008) AcsD catalyzes enantioselective citrate desymmetrization in siderophore biosynthesis. Nature Chem Biol 5:174–182

Schwyn B, Neilands JB (1987) Universal chemical assay for the detection and determination of siderophores. Anal Biochem 160:47–56

Segond D, Dellagi A, Lanquar V, Rigault M, Patrit O, Thomine S, Expert D (2009) NRAMP genes function in *Arabidopsis thaliana* resistance to *Erwinia chrysanthemi* infection. Plant J 58:195–207

Silby MW, Winstanley C, Godfrey SAC, Levy SB, Jackson RW (2011) *Pseudomonas* genomes: diverse and adaptable. FEMS Microbiol Rev 35:652–680

Smits TH, Duffy B (2011) Genomics of iron acquisition in the plant pathogen *Erwinia amylovora*: insights in the biosynthetic pathway of the siderophore desferrioxamine E. Arch Microbiol 193:693–639

Sonoda H, Suzuki K, Yoshida K (2002) Gene cluster for ferric uptake in *Agrobacterium tumefaciens* MAFF301001. Genes Genet Syst 77:137–146

Subramoni S, Sonti RV (2005) Growth deficiency of a *Xanthomonas oryzae* pv. *oryzae fur* mutant in rice leaves is rescued by ascorbic acid supplementation. Mol Plant-Microbe Interact 18:644–651

Swingle B, Thete D, Moll M, Myers CR, Schneider DJ, Cartinhour S (2008) Characterization of the PvdS-regulated promoter motif in *Pseudomonas syringae* pv. *tomato* DC3000 reveals regulon members and insights regarding PvdS function in other pseudomonads. Mol Microbiol 68:871–889

Taguchi F, Suzuki T, Inagaki Y, Toyoda K, Shiraishi T, Ichinose Y (2010) The siderophore pyoverdine of *Pseudomonas syringae* pv. *tabaci* 6605 is an intrinsic virulence factor in host tobacco infection. J Bacteriol 192:117–126

Tomisić V, Blanc S, Elhabiri M, Expert D, Albrecht-Gary AM (2008) Iron(III) uptake and release by chrysobactin, a siderophore of the phytophatogenic bacterium *Erwinia chrysanthemi*. Inorg Chem 47:9419–9430

Turgeon BG, Baker SC (2007) Genetic and genomic dissection of Cochliobolus heterostrophus *Tox1* locus controlling biosynthesis of the polyketide virulence factor T-toxin. Adv Genet 57:219–261

Visca P, Leoni L, Wilson MJ, Lamont IL (2002) Iron transport and regulation, cell signalling and genomics: lessons from *Escherichia coli* and *Pseudomonas*. Mol Microbiol 45:1177–1790

Wensing A, Braun SD, Büttner P, Expert D, Völksch B, Ullrich MS, Weingart H (2010) Impact of siderophore production by Pseudomonas syringae pv. syringae 22d/93 on epiphytic fitness and biocontrol activity against Pseudomonas syringae pv. glycinea 1a/96. Appl Environ Microbiol 76:2704–2011

Wiggerich H-G, Pühler A (2000) The *exbD2* gene as well as the iron-uptake genes *tonB*, *exbB* and *exbD1* of *Xanthomonas campestris* pv. *campestris* are essential for the induction of a hypersensitive response on pepper. Microbiology 146:1053–1060

Winkelmann G (2007) Ecology of siderophores with special reference to the fungi. Biometals 20:379–392

Yang S, Perna NT, Cooksey DA, Okinaka Y, Lindow SE, Ibekwe AM, Keen NT, Yang CH (2004) Genome-wide identification of plant upregulated genes of *Erwinia chrysanthemi* 3937 using a GFP-based IVET leaf array. Mol Plant Microbe Interact 17:999–1008

Yokosho K, Yamaji N, Ueno D, Mitani N, Ma JF (2009) OsFRDL1 is a citarte transporter required for efficient translocation of iron in rice. Plant Physiol 149:297–305

Zaini PA, Fogaça AC, Lupo FGN, Nakaya HI, vencio RZN, da Silva AM (2008) The iron stimulon of *Xylella fastidiosa* includes genes for Type IV pilus and colicin V-like bacteriocins. J Bacteriol 190:2368–2378

Zhao Y, Blumer SE, Sundin GW (2005) Identification of *Erwinia amylovora* genes induced during infection of immature pear tissue. J Bacteriol 187:8088–8103

Chapter 3
Mechanisms and Regulation of Iron Homeostasis in the Rhizobia

Elena Fabiano and Mark R. O'Brian

Abstract Rhizobia are soil bacteria belonging to different genera whose most conspicuous characteristic is the ability to establish a symbiotic association with legumes and carry out nitrogen fixation. The success of these organisms in the rhizosphere or within the host plant involves the ability to sense the environment to assess the availability of nutrients, and to optimize cellular systems for their acquisition. Iron in natural habitats is mostly inaccessible due to low solubility, and microorganisms must compete for this limited nutrient. In addition to their agricultural and economic importance, rhizobia are model organisms that have given new insights into related, but less tractable animal pathogens. In particular, genetic control of iron homeostasis in the rhizobia and other α-Proteobacteria has moved away from the Fur paradigm to an iron sensing mechanism responding to the metal indirectly. Moreover, utilization of heme as an iron source is not unique to animal pathogens, and the rhizobial strategy reveals some interesting novel features. This chapter reviews advances in our understanding of iron metabolism in rhizobia.

Keywords Iron acquisition · Heme · Rhizobial–legume interactions · Metalloregulation

Rhizobia are a diverse group of Gram-negative soil bacteria that can form a symbiosis with leguminous plants. In general, a rhizobial species recognizes one or

E. Fabiano
Departmento de Bioquímica y Genómica Microbianas, Instituto de Investigaciones Biológicas Clemente Estable, Av. Italia 3318, 11600, Montevideo, Uruguay
e-mail: efabiano@iibce.edu.uy

M. R. O'Brian (✉)
Department of Biochemistry, State University of New York at Buffalo,
140 Farber Hall, Buffalo, NY 14214, USA
e-mail: mrobrian@buffalo.edu

D. Expert and M. R. O'Brian (eds.), *Molecular Aspects of Iron Metabolism in Pathogenic and Symbiotic Plant–Microbe Associations*, SpringerBriefs in Biometals,
DOI: 10.1007/978-94-007-5267-2_3, © The Author(s) 2012

a few legume hosts; the molecular basis for the specificity between the bacterium and host, and the events leading to the symbiotic state are understood in considerable detail (Downie and Walker 1999; Oke and Long 1999; Spaink 2000; Gibson et al. 2008; Oldroyd and Downie 2008; Masson-Boivin et al. 2009). Rhizobia elicit the formation of a symbiotic organ called a root nodule comprising differentiated plant and bacterial cells. In this context, rhizobia are endosymbionts within the plant cells of the nodule.

Differentiated rhizobia within nodules are termed bacteroids, and acquire the ability to convert atmospheric nitrogen into ammonia by a process called nitrogen fixation. Nitrogen fixed by the endosymbiont is exported out of the cell in the form of ammonium or amino acids, and subsequently assimilated by the plant host to fulfill its nutritional nitrogen requirement. In return, the plant provides the bacteroids with a carbon source ultimately derived from photosynthesis. The rhizobia-legume symbioses account for nearly half of the global biological nitrogen fixation (Gruber and Galloway 2008), and is therefore an important agricultural and environmental process. Moreover, the rhizobia are phylogenetically related to several plant and animal pathogens, and serve as a model to understand the molecular basis of bacterial–eukaryote interactions.

Rhizobia include the genera *Rhizobium*, *Bradyrhizobium*, *Sinorhizobium*, *Mesorhizobium*, *Azorhizobium*, and *Allorhizobium*. These genera belong to the α-Proteobacterial subdivision of the purple bacteria, an extremely diverse group that includes pathogens, symbionts, photosynthetic organisms, bacteria that degrade environmental pollutants, and the abundant marine organism *Pelagibacter ubique* (Ettema and Andersson 2009). The bacterial ancestor of mitochondria belongs to this group as well. More recently, numerous α-Proteobacterial species have also been identified that form symbiosis with legumes (Moulin et al. 2001; Chen et al. 2003; Taulé et al. 2012). Although these bacteria are also referred to as rhizobia, they are phylogenetically distinct from the α-Proteobacterial species, and are not considered further here.

In recent years, studies on rhizobial iron metabolism have focused to a large extent on three species, *Rhizobium leguminosarum*, *Sinorhizobium meliloti* and *Bradyrhizobium japonicum*. *R. leguminosarum* and *S. meliloti* belong to the family Rhizobiaceae, whereas *Bradyrhizobium* belongs to the *Bradyrhizobiaceae*. These two families diverged approximately 500 million years ago (Turner and Young 2000), and thus it is not unexpected that important differences between them have been observed with respect to iron metabolism.

3.1 The Problem of Iron Acquisition and Roles of Iron in Symbiosis

As a free-living organism, rhizobia must compete with other soil microbiota for nutrients and other resources, including iron. Iron is an abundant element, but it is primarily in the oxidized form, which has an extremely low solubility in aqueous environments at neutral pH. As described below, bacteria have developed numerous strategies to acquire iron from the environment, including highjacking other microbial iron acquisition systems. At present, much more is known about iron acquisition and regulation in free-living rhizobia than in symbiosis.

Bacteroids within a functional nodule are surrounded by the peribacteroid membrane, a plant structure which separates the bacteria from the host within the plant cell. The peribacteroid membrane encompasses one or many bacteroids, depending on the type of nodule formed by the plant, and this structure is called a symbiosome. This means that the supply of iron and other nutrients to the bacteria is ultimately controlled by the plant. Whereas bacteria in the rhizosphere likely experience iron stress due to the low solubility of ferric iron, it cannot be assumed that the symbiotic prokaryote is iron-limited despite the large demand. At present, the mode of transport of iron into bacteroids is not known, nor is it known whether a high affinity transporter is necessary. Evidence for an iron pool in the peribacteroid space of soybean nodules could suggest that the bacteroids perceive a high iron environment (LeVier et al. 1996), but none of these issues have been resolved.

Iron is a major constituent of two crucial proteins for the N_2-fixing symbiosis process, bacterial nitrogenase and plant leghemoglobin. Moreover, other iron containing proteins which contribute to the N_2 fixation process are induced symbiotically, such as rhizobial hydrogenase and cytochromes that allow respiration in the low oxygen milieu of the nodule (Hennecke 1992; Sangwan and O'Brian 1992; Baginsky et al. 2002).

The nitrogenase complex contains about 34 atoms of iron. It comprises the iron protein (the nucleotide binding and electron-donating element) and the iron–molybdenum protein which contains the N_2-reducing site. The iron protein is a homodimer of two subunits that coordinate one [4Fe-4S] cluster. The iron–molybdenum protein is a tetramer of two polypeptides ($\alpha_2\beta_2$), where the α subunit contains a [Mo-7Fe-9S-homocitrate] cluster and the β subunit a [8Fe-7S] cluster (Georgiadis et al. 1992; Rees and Howard 2000). Nitrogenase constitutes about 30 % of the soluble proteins (11 % of total proteins) in the bacteroid (Verma and Nadler 1984).

In N_2-fixing nodules, leghemoglobin can represent about 20 % of the total nodule iron (1–3 mg of leghemoglobin per gram of fresh weight depending on plant species) (Dilworth 1980). As previously mentioned, differentiated bacteroids are enveloped in a peribacteroid membrane and immersed in a leghemoglobin-rich environment. It is interesting to note that peribacteroid membrane is one of the first structures to be degraded during the nodule senescence (Herrada et al. 1993). In this scenario, it is possible that iron requirement is greater for nodulated plants than for host plants alone. In fact, it has been reported that iron deficiency negatively affects the rhizobia–legume symbiosis. Application of the metal to iron-stressed plants increases nodule number and plant mass. Furthermore, S. meliloti iron-starved cells are less competitive for nodulation, and iron deficiency prevents nodule initiation and limits nodule development (O'Hara et al. 1988; Tang et al. 1990; Expert and Gill 1992; Battistoni et al. 2002b).

3.2 Iron Transport Systems

3.2.1 Siderophore-Mediated Iron Transport

A common strategy used by many bacteria to obtain iron from their environment when it is scarce involves the production and secretion of ferric chelating compounds termed siderophores (Greek for iron carriers) (Lankford 1973; Neilands 1973, 1981; Ratledge and Dover 2000). Siderophores are low molecular weight compounds (around 300–1,500 Da) that bind iron specifically and with high affinity. The secreted siderophore, bound to iron, is then taken up by the cell through a ferric siderophore transport system. In Gram-negative bacteria, translocation of ferric siderophore complexes across the outer membrane requires energized transport via outer membrane receptors. The energy is provided by the proton motive force, and it is transduced to outer membrane receptors by the TonB-ExbB-ExbD complex (Postle and Kadner 2003).

Although siderophores differ widely in their overall structure, variation in metal-binding groups is more limited. According to the iron-chelating group, siderophores are classified as catecholates, hydroxamates, α-hydroxycarboxylates, or mixed siderophores. Although these are the most common groups, other metalbinding groups have been identified, for example oxazoline, thiazoline, hydroxypyridine, and β-hydroxyacids. Siderophores can also be characterized according to their biosynthetic mechanism as nonribosomal peptide synthetase (NRPS)-dependent or NRPS-independent (Raymond and Dertz 2004; Challis 2005; Donadio et al. 2007).

Three rhizobial endogenous siderophores have been chemically characterized to date (Fig. 3.1): (i) Vicibactins, trihydroxamate NRPS-dependent siderophores synthetized by R. leguminosarum. (ii) Rhizobactin, and α-hydroxycarboxylate siderophore produced by S. meliloti strain DM1 (iii) Rhizobactin 1021, a citrate-derivative di-hydoxamate and NRPS-independent siderophore syntethized by S. meliloti strain 1021 (Persmark et al. 1993; Dilworth et al. 1998; Carter et al. 2002; Smith et al. 1985). Siderophore biosynthesis has been studied in R. leguminosarum 8401(pRL1JI) and S. meliloti 1021. In both cases, biosynthetic genes are located on plasmids and are clustered close to their respective outer membrane receptors (Fig. 3.2). In addition, some R. leguminosarum species produce trihydroxymates, while some sinorhizobia produce dihydroxymates or α-hydroxycarboxylates although chemical structures for them are unavailable (Carson et al. 2000). Catechol-like siderophore production has been reported in R. leguminosarum IARI 102 and Rhizobium sp (cowpea) RA-1 (Modi et al. 1985; Patel et al. 1988). Unfortunately, none of these structures have been resolved. Anthranilate, a precursor of tryptophan, has been reported to promote the uptake of ferric iron into iron-starved cells in R. leguminosarum (Rioux et al. 1986). Anthranilate was synthesized and transported in cultures grown with iron, and therefore is not a specific response to iron deficiency. Anthranilate synthesis mutants of S. meliloti form defective nodules that cannot be accounted for by a tryptophan auxotrophy

Fig. 3.1 Chemical structures of three siderophores characterized in rhizobia

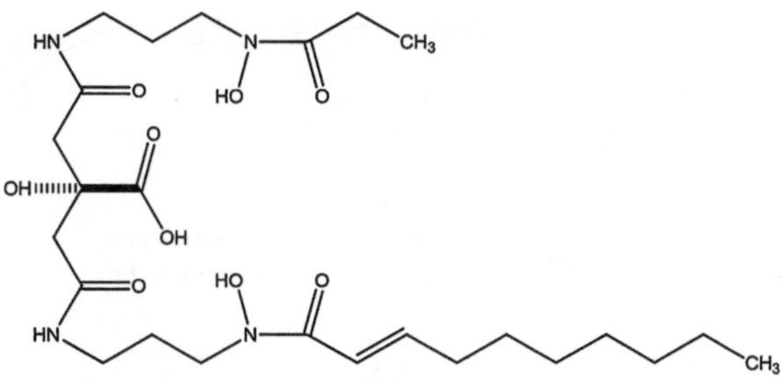

Rhizobactin 1021. The siderophore produced by *S. meliloti* 1021.

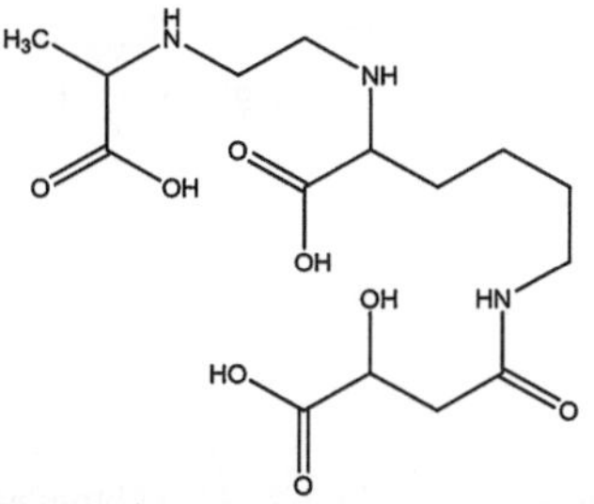

Rhizobactin. The siderophore produced by *S. meliloti* DM4

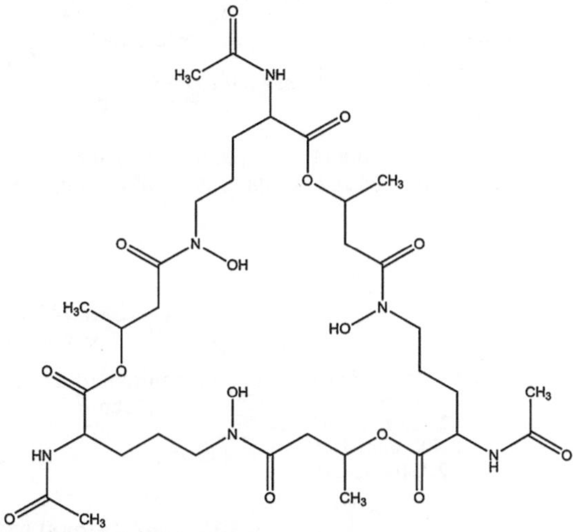

Vicibactin. The siderophore produced by *R. leguminosarum* WSM710

since downstream mutants in the tryptophan pathway form effective nodules (Barsomian et al. 1992). Interestingly, the *B. japonicum* 110 genome is a single chromosome with no plasmids, and no endogenous siderophore has been identified in that species. Nevertheless, *B. japonicum* does have genes encoding siderophore receptors, as discussed below.

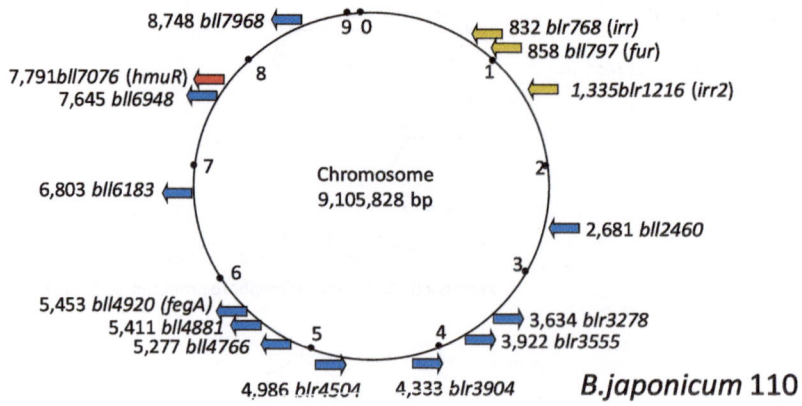

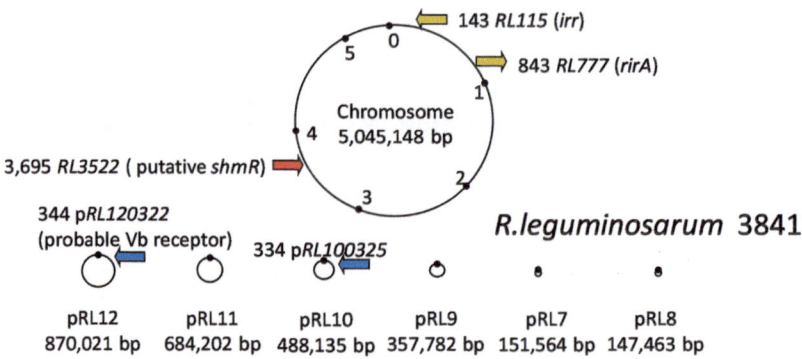

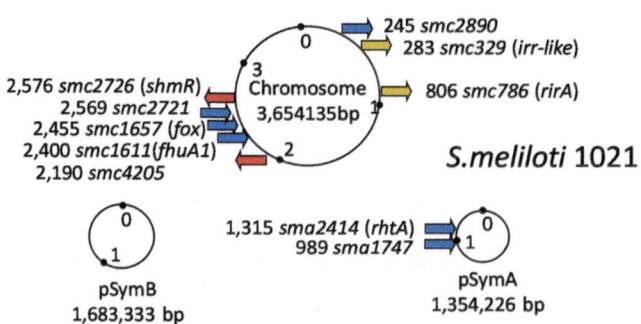

Fig. 3.2 Genetic maps of genes coding putative TonB-dependent receptors and iron regulators in *B. japonicum* 110, *R. leguminosarum* 3841, and *S. meliloti* 1021. Genes coding heme receptors are shown as *gray arrows*, other putative TonB-dependent receptor genes as *black arrows*, and iron regulator genes as empty *arrows*. The directions of transcription are indicated by the *arrows*. The distances in Mbp from the origins of replication are indicated as *dots*. The approximate locations of the genes are shown in kbp before the name of the genes. Sizes of chromosomes are indicated into the *circles*, and sizes and names of plasmids at the *bottom* of each plasmid

3.2.1.1 Vicibactin

Vicibactin, the siderophore made by *R. leguminosarum*, is a cyclic trihydroxamate containing three residues each of D-3-hydroxybutanoic acid and *N*2-acetyl-*N*5-hydroxy-D-ornithine arranged in alternate ester and peptide bonds (Dilworth et al. 1998) (Fig. 3.1). Vicibactin was identified from three different *R. leguminosarum* strains, WSM709, WSM710, and WU235. The sequenced *R. leguminosarum* 3841 strain also contains genes presumably involved in vicibactin biosynthesis located on plasmid pRL12 and close to the putative vicibactin receptor (Fig. 3.2).

Vicibactin biosynthesis involves an NRPS gene homolog, *vbsS* (Carter et al. 2002). NRPSs are assembly lines of specialized domains that form peptide bonds independently of ribosomes (Finking and Marahiel 2004; Donadio et al. 2007). NRPSs use amino or hydroxy-acids as building blocks, catalyzing the formation of amide or ester bonds, respectively. Each NRPS module consists of three basic domains: an adenylation domain (A) that activates the selected amino acid as an acyl adenylate, a peptide carrier (PCP) or T domain responsible for the formation of the thioester bond, and a condensation domain (C) which catalyzes peptide bond formation between two activated amino acids elongating the peptide and translocating it to the PCP domain. These three domains form a module, and NRPSs are consecutive tandems of these modules with each module being responsible for the addition of a new amino acid. Modules are colinear with the amino acids they link. A final module containing a thioesterase (TE) domain, instead of the C domain, completes the peptide formation. Additional activities may be catalyzed by specialized domains such as cyclization (Cy domain), epimerization (E domain), methylation (M domain), or oxidation or reduction (Finking and Marahiel 2004; Donadio et al. 2007).

Biosynthesis of vicibactin involves a cluster of eight genes arranged in four operons, *vbsGSO*, *vbsADL*, *vbsC*, and *vbsP* (Dilworth et al. 1998; Carter et al. 2002). A biosynthetic pathway has been proposed based on some mutant phenotypes and in silico studies. According to the model, the first step in biosynthesis is the hydroxylation of L-ornithine to give LGENES ARE MAXIMALLY EXPRESSED-N^5-hydroxyornithine probably catalyzed by VbsO. L-N^5-hydroxyornithine is then the substrate for VbsS, an enzyme similar to NRPS. VbsS activates the substrate by adenylation and transfers it to the PCP domain. The location and sequence of *vbsP* suggests that VbsP is responsible for the addition of phosphopantotheinate to the PCP domain, although this gene is not essential probably due to the presence of another phosphopantetheinyl transferase. The addition of butyryl coenzyme A to hydroxyornithine attached to VbsS, may be then catalyzed by VbsA. The resulting N^5-hydroxy-N^5-(D-3-hydroxybutyryl)-ornithine could be subsequently epimerized by VbsL. After the acetylation catalyzed by VbsC, the N^2-acetyl-N^5-hydroxy-N^5-(D-3- hydroxybutyryl)-ornithine bound as a thioester to the PCP domain of VbsS may be trimerized and cyclized by the TE domain of VbsS. The *vbsG* gene is essential for vicibactin biosynthesis although its function is still unknown. Based on sequence similarity, it has been suggested that the *vbsD* gene could be involved in siderophore export, although it is not essential in *R. leguminosarum* as *vbsD* mutants that accumulate normal amounts of siderophores (Yeoman et al. 2000). Vicibactin secretion is a topic that requires further investigation.

In *R. leguminosarum* 8401(pRL1JI) a gene, *fhuA1*, similar to the ferrichrome receptor gene from *E.coli,* is found adjacent to the *vbs* genes (Yeoman et al. 2000). In *E. coli*, the ferrichrome uptake system is encoded by the *fhuACDB* operon. FhuA is the outer membrane receptor, while FhuBCD is the ABC transport system involved in ferrichrome uptake: FhuB is a permease, FhuC an ATP-binding cassette protein, and FhuD the periplasmic protein (Burkhardt and Braun 1987; Coulton et al. 1987; Koster and Braun 1990). Upstream and transcribed in the same orientation as *fhuA1*, is located a gene, *fhuF,* whose predicted protein is similar to the *E. coli* putative ferric reductase. It has been suggested that it may be involved in iron release from the ferric-vicibactin complex. The *fhuCDB* operon is predicted to encode a periplasmic vicibactin-binding protein (FhuD), an inner membrane transporter (FhuB), and an ATPase (FhuC) involved in vicibactin transport through the inner membrane (Stevens et al. 1999). Vicibactin internalization requires a functional TonB protein (Wexler et al. 2001).

3.2.1.2 Rhizobactin 1021

The siderophore produced by *S. meliloti* 1021 was named rhizobactin 1021 to differentiate it from rhizobactin, an amino polycarboxylic acid siderophore produced by *S. meliloti* DM4 (Persmark et al. 1993; Smith et al. 1985) (Fig. 3.1). Rhizobactin was the first siderophore produced by *S. meliloti* to be structurally characterized, but production of rhizobactin 1021 seems to be more widespread among sinorhizobial strains. Rhizobactin 1021 is an asymmetrical citrate-derivative hydroxamate carrying two different acyl groups, acetyl and 2-decenoyl. Other citrate-hydroxamate siderophores are: schizokinen produced by some *Bacillus* species and cyanobacteria, aerobactin from enteric bacteria, arthrobactin from *Arthrobacter pascens*, acinetoferrin from *Acinetobacter haemolyticus,* and nannochelin A from *Nannocystis exedens* (Williams 1979; Neilands and Leong 1986; Kunze et al. 1992; Okujo et al. 1994; Wilhelm and Trick 1994; Hu and Boyer 1995). Citrate-hydroxamate siderophores represent two types of ferric ligands in one molecule, since the ferric iron is coordinated with two hydroxamates, and the 2-hydroxycarboxylate group from the citric acid.

In *S. meliloti* 1021, genes required for rhizobactin 1021 synthesis are located in the pSymA plasmid (symbiotic plasmid A) which also contains genes for nodulation (*nod* genes) and nitrogen fixation (*nif* genes). Rhizobactin biosynthesis involves at least a cluster of six genes arranged in one operon, *rhbABCDEF* (Lynch et al. 2001). A plausible biosynthetic pathway was proposed on the basis of gene disruption, predicted function of gene products, and the chemical structure of rhizobactin 1021. It was suggested that the first step involves the synthesis of 1,3-diaminopropane. This compound is produced from L-glutamic acid and L-aspartic-B-semialdehyde by a two-step reaction, catalyzed first by RhbA to produce L-2,4-diamino butyric acid and 2-ketoglutaric acid and then by RhbB, responsible for decarboxylation of L-2,4-diamino butyric acid. The 1,3-diaminopropane is then hydroxylated by RhbE to produce N^4-hydroxy-1-aminopropane which is in turn acetylated by RhbD to form N^4-acetyl-N^4-hydroxy-1-aminopropane. The coupling of two molecules of N^4-acetyl-N^4-hydroxy-1-aminopropane to citrate may occur in a similar way as aerobactin production in *E. coli*. By analogy to aerobactin synthesis, it is proposed that this reaction consists of two steps catalyzed by a synthetase complex of two subunits encoded by RhbC and RhbF.

One step of the rhizobactin 1021 synthesis pathway requires an N-acylase for addition of the lipid moiety. RhbG is similar to some acyl-CoA-dependent acyl transferases and therefore it seems likely to be responsible for acylation with 2-decenoyl-CoA. Lynch et al. (2001) propose that acylation could be the last step of the pathway, nonetheless Challis (2005), proposes a modified biosynthetic pathway, where N^4-hydroxy-1-aminopropane could be acetylated with acetyl-CoA through RhbD or alternatively with 2-decenoyl-CoA through RhbG. Disruption of the *rhbG* gene did not affect siderophore production or uptake as determined by chrome azurol S (CAS) or bioassays, respectively. However, it could not be discounted that the nonlipidic derivative is still used as a siderophore (Lynch et al. 2001). In summary, the role of RhbG in rhizobactin 1021 biosynthesis and how the lipid moiety is incorporated, await further investigations. The secretion system used by rhizobactin 1021 is also an open question.

The outer membrane receptor for rhizobactin 1021 is encoded by the *rhtA* gene, localized upstream of the siderophore biosynthetic genes but transcribed in the opposite orientation. The RhtA amino acid sequence reveals a TonB box and is homologous with IutA, the aerobactin receptor of *E. coli* (Lynch et al. 2001).

Upstream of the *rhbABCDEF* genes and in the same operon is localized the *rhtX* gene. Interestingly, disruption of *rhtX* gene abolishes rhizobactin 1021 utilization. Moreover, RhtX (a member of the Major Facilitator Superfamily of solute transporters) alone could substitute for FhuCDB in rhizobactin 1021 transport in *E. coli*. On these bases, Cuiv et al. (2004) propose RhtX as a permease that belongs to a novel family of siderophore transport systems not related to the ABC family of transporters (Fig. 3.3).

3.2.1.3 Xenosiderophore Transport

Soil is an important niche for many diverse genera of microbes, and rhizobia must compete with these organisms for nutritional iron in the rhizosphere. One strategy employed by soil microbes for iron acquisition is to take up ferric siderophores synthesized and secreted by other organisms. These are referred to as xenosiderophores (or exogenous siderophores). In the upper layer of humic soil, streptomycetes and fungi are predominant microorganisms, and they produce the trihydroxamate siderophores desferrioxamines and desferrichrome, respectively (Winkelmann 2007). Ferrichrome and ferrioxamines can be used as iron sources by many fungi and by some Gram-positive and Gram-negative bacteria as is the case for some rhizobia strains. The use of xenosidephores was found to be a variable trait in rhizobia. It has been reported that ferrichrome and rhodotorulic acid (a trihydroxamate siderophore produced by some yeasts) stimulate the growth

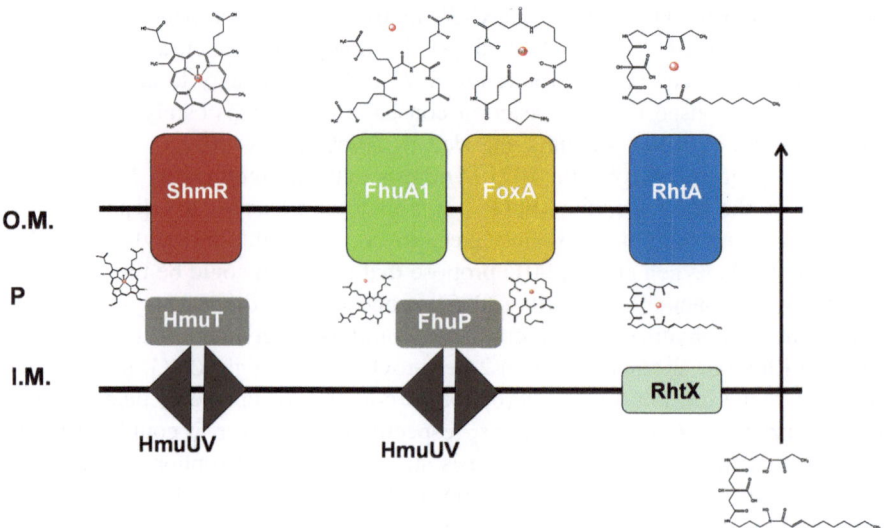

Fig. 3.3 Model of iron acquisition systems expressed in *S. meliloti*. Heme transport system involves ShmR, the outer membrane heme receptor; HmuT, a periplasmic binding protein; HmuU, a membrane permease, and HmuV, an ATPase. Ferrichrome and ferrioxamine B transport systems consist of FhuA1, the ferrichrome outer membrane receptor and FoxA, the ferrioxamine B outer membrane receptor; FhuP, a periplasmic binding protein for both siderophores and the HmuUV transporter. The ferric rhizobactin1021 transport system requires RhtA, the siderophore outer membrane receptor and RhtX, a permease that belong to a novel family of siderophore transporters. Abbreviations: *O.M.*, outer membrane; *P*, perisplasm; *I.M.*, inner membrane

of *B. japonicum* 110, LO, and 61A152 strains (Plessner et al. 1993; Small et al. 2009; Small and O'Brian 2011). Some, but not all, sinorhizobia strains tested are able to use ferrichrome, ferric rhodotorulate, and ferrioxamine B as iron sources (Smith and Neilands 1984; Reigh and O'Connell 1993). One out of eight *R. leguminosarum* strains tested can utilize ferrioxamine B, while the remaining strains could not (Skorupska et al. 1989). Only 3 out of 17 *Mesorhizobium* strains tested used ferrichrome as a sole iron source (Carlton et al. 2007). To explain the variation in the ability of using ferrichrome as iron source in *Mesorhizobium* strains, Carlton et al. (2007) suggest that ferrichrome transport systems evolved through cycles of gene acquisition and deletion, with the positive selection pressure of an iron-poor or siderophore-rich environment.

There are several reports of xenosiderophores use by rhizobia, but only a few are well characterized. Inspection of the genomes of *B. japonicum* 110, *R. leguminosarum* 3841, and *S. meliloti* 1021 reveals the presence of numerous putative TonB-dependent receptors (Fig. 3.2). In addition to endogenous siderophore receptors and putative heme receptors, the number of TonB-dependent receptors differs in these three rhizobia strains; there is only one in *R. leguminosarum* 3841, five in *S. meliloti* 1021, and eleven in *B. japonicum* 110. As only a few of them have been characterized, it cannot be assumed that they are all involved in iron transport. In fact, two *B. japonicum* TonB-dependent receptor genes, *bll4766* and *bll6848*, have been suggested to be involved in cobalamin and nickel transport, respectively (Schauer et al. 2008). Based on growth assays, iron uptake experiments, and *in silico* analysis, no obvious receptors for catechol-like siderophore have been identified in the rhizobia, while ferrichrome seems to be the xenosiderophore more widely used.

Five putative TonB-dependent outer membrane receptor genes in *B. japonicum* strains 110 and LO are strongly induced by iron limitation (Yang et al. 2006b; Small et al. 2009b), and thus are the best candidates to encode receptors for iron siderophores. One of them, *bll4920*, is highly similar to the ferrichrome receptor *fegA* gene identified previously for *B. japonicum* strain 61A152 (LeVier et al. 1996; Benson et al. 2005), and is essential for ferrichrome utilization (Small et al. 2009b). The *fhuE* (*blr4504*) and pyoR (blr3555) genes encode outer membrane receptors for rhodotorulic acid and pyoverdine PL-8, respectively based on growth phenotypes of mutants defective in those genes (Small and O'Brian 2011).

None of the five putative outer membrane receptor genes in *B. japonicum* are proximal to other ABC transporter genes necessary for uptake into the cytoplasm as is found in *E. coli*, and thus no ferric iron transport system has been completely characterized in that species. The *fegA* gene in strain 61A152 is found in an operon with *fegB* located downstream of it (Benson et al. 2005), but the *fegB* gene is not found in strains 110 or LO (Small et al. 2009b). The function of *fegB* is unknown.

O'Connell and colleagues have revealed a novel route of ferrichrome and ferrioxamine B transport in *S. meliloti* (Cuiv et al. 2008). In *S. meliloti* 2011, a strain closely related to the sequenced *S. meliloti* 1021 strain, ferrichrome and ferrioxamine B outer membrane receptors are encoded by homologs of *smc01611* and *smc01657*. These genes were termed *fhuA1* and *foxA*, respectively, by analogy to enteric bacteria. A third *fhuA* homolog, *sma1747*, is also found (Fig. 3.2), although it is not involved in ferrichrome or ferrioxamine B uptake, and its role remains unknown. A predicted periplasmic protein encoded by the *fhuP* gene localized downstream the *foxA* gene was shown to be involved in both ferrichrome and ferrioxamine B uptake. Intriguingly, *hmuU* and *hmuV* mutants were impaired in utilization of these xenosiderophores as iron sources. The ABC system *hmuTUV* has been shown to be necessary for hemin and hemoglobin utilization in rhizobia (Fig. 3.3) (Cuiv et al. 2008).The ability of one permease complex, *hmuUV*, to transport two structurally unrelated compounds, hydroxamate siderophores and heme, is unusual. From this, it follows that more than one periplasmic binding protein should be able to interact with a single inner membrane permease, which allows the delivery of both siderophore and heme to the cytoplasm. Alternatively, iron may be released from iron compounds in the periplasm, and then transported by a common permease system. It has been proposed that in *Pseudomonas aeruginosa*, iron may be removed from some siderophores in the periplasm and that an iron-specific rather than a siderophore-specific transporter delivers the metal to the cytoplasm (Poole 2004). Whereas a ferric reductase activity may be sufficient to remove iron complexed to a siderophore, reduction of iron from heme does not release the metal, and thus a periplasmic heme-degrading enzyme would be necessary (see below). From this hmuUV likely transports two different substrates.

3.2.1.4 Ferric Iron Reduction

Once ferric chelates are taken up into cells, the iron must be reduced to the ferrous form, but the proteins responsible for this activity have only been identified in a few cases (Matzanke et al. 2004; Sedlacek et al. 2009; Wang et al. 2011). Most bacterial assimilatory ferric reductases are soluble and require flavin for activity. An exception is FhuF, an iron–sulfur cytoplasmic protein that reduces ferric ferrioxamine B (Matzanke et al. 2004).

The *B. japonicum frcB* (bll3557) gene was identified as a gene adjacent to, and co-regulated with, the *pyoR* gene encoding the receptor for ferric-pyoverdine (Small and O'Brian, 2011). FrcB is a membrane-bound, di-heme protein, characteristics of eukaryotic ferric reductases (Dancis et al. 1990; Robinson et al. 1999; McKie 2008). Heme is essential for FrcB stability, as were conserved histidine residues in the protein that likely coordinate the heme moieties. Expression of the *frcB* gene in *E. coli* conferred ferric reductase activity on those cells. Furthermore, purified, reduced FrcB was oxidized by ferric iron in vitro. *B. japonicum* cells showed inducible ferric reductase activity in iron-limited cells that was diminished in an *frcB* mutant. Steady-state levels of *frcB* mRNA was strongly induced under iron limitation in an Irr-dependent manner. FrcB belongs to a family of previously uncharacterized proteins found in many Proteobacteria and some cyanobacteria. This suggests that membrane-bound, heme-containing ferric reductase proteins are not confined to eukaryotes, but rather they may be common in bacteria.

3.2.2 Heme and Heme Proteins as Iron Sources

Heme is an iron-containing porphyrin compound essential for many cellular processes including oxygen transport, energy generation, cellular redox reactions, and metabolic regulation (O'Brian and Thony-Meyer 2002). It can also be a source of iron to bacteria that have access to it. Bacterial pathogens of animals can use hemoglobin and other heme proteins as iron sources (Wandersman and Stojiljkovic 2000; Genco and Dixon 2001; Wandersman and Delepelaire 2004). The concentration of free heme is usually very low, as it is bound tightly to hemoproteins or sequestered by serum albumin or hemopexin. Similar to siderophore transport, the use of heme or heme compounds relies on TonB-dependent outer membrane receptors able to transport heme, but not the apoprotein, into the periplasm and on ABC transporters to internalize it into the cytosol. Heme transport systems are maximally expressed when iron is scarce. Heme receptor proteins can be classified into two families. The first family comprises receptors that directly interact with free heme or hemoproteins. The second family includes receptors that depend on extracellular heme binding proteins called hemophores (Wandersman and Delepelaire 2004). Hemophores are proteins produced and secreted by the bacterium that sequester heme from diverse sources and then binds to an outer membrane hemophore receptor, followed by transport into the periplasm. Once in the periplasm, heme is then transported into the cytoplasm by an ABC transporter. The Hem system of *Yersinia enterocolitica* (HemTUV) was among the first ABC transporters involved in heme uptake to be characterized, and is considered as a prototype for heme transport. Its homolog in *Y. pestis* is called as HmuTUV. According to the model of Hmu transport, once the heme moiety is translocated into the periplasm it is bound by a periplasmic binding protein HmuT which in turn presents it to the inner membrane permease-ATP hydrolase complex of HmuU/HmuV. Intact heme is then delivered to the bacterial cytoplasm. Interestingly, mutations in heme-specific ABC transport systems do not always result in abrogation of heme utilization by the bacterium. Recently, an alternative inner membrane system involved in heme uptake was discovered by Wandersman and co-workers (Letoffe et al. 2006, 2008), who found that the dipeptide periplasmic binding protein DppA or the L-alanyl-γ-D-glutamyl-meso diaminopimelate periplasmic binding protein MppA, are implicated in heme transport together with the dipeptide permease ABC, DppBCDF complex.

For many years, transport and degradation of heme as a mechanism to acquire iron was thought to be exclusive to pathogens. However, Noya et al. (1997) discovered that rhizobia and other non-pathogenic bacteria can use heme or hemoglobin as iron sources. Subsequently, genes involved in heme transport were identified in *B. japonicum*, *R. leguminosarum,* and *S. meliloti* (Nienaber et al. 2001; Wexler et al. 2001; Battistoni et al. 2002a; Amarelle et al. 2008). Notably, while the use of xenosiderophores was found to be variable, heme transport systems are a common trait of rhizobia strains studied so far. Moreover, rhizobia heme transport systems are present in the chromosome and not in plasmids (Fig. 3.2). As mentioned before, leghemoglobin is the most abundant protein in the cytosol of symbiotic root nodule. Although rhizobia are not in direct contact with leghemoglobin within the nodule cytosol, they may have access to it during nodule senescence. The possibility of leghemoglobin as an iron source under physiological conditions has not been addressed experimentally.

3.2.2.1 Heme Transport

The *B. japonicum bll07076* gene and the *S. meliloti smc02726* gene coding for heme receptor proteins HmuR and ShmR respectively, are required for utilization of heme, hemoglobin, and leghemoglobin. Both predicted proteins are similar to outer membrane TonB-dependent receptors, and display the typical FRAP/NPNL motif of heme receptors, although the highly conserved histidine residue in the FRAP/NPNL region is substituted by asparagine in ShmR. Despite the similarity of these proteins with other heme receptors, the identity of these two rhizobial proteins to each other is only 23 %. In *R. leguminosarum,* a predicted protein coded by *rll3522* and showing 65 % identity with ShmR was identified by in silico studies, even though its function in heme uptake has not yet been described. Analysis of the *S. meliloti* 1021 genome also reveals the presence of a second putative heme receptor protein: Smc04205. The predicted sequence of Smc04205 indicates that this protein is homologous to HasR-like outer membrane receptors which recognize heme-bound hemophores. Moreover, the genetic region around *smc04205* is syntenous with the *hasR* regions of numerous other bacteria (Cescau et al. 2007). Nonetheless, no hemophore-like protein has yet been detected in rhizobia and no *hasR* promoter activity could be found.

In *B. japonicum, hmuR* is proximal and divergently transcribed from *hmuPTUV* genes. In *S. meliloti* and in *R. leguminosarum, hmuTUV* genes are distal from *shmR*. By heterologous expression in *E. coli*, Cuiv et al. (2008) confirmed that *S. meliloti hmuTUV* genes are implicated in heme transport in *E. coli*. Intriguingly, *hmuTUV* mutants in *B. japonicum, R. leguminosarum,* and *S. meliloti* presented a reduced but not abolished use of heme and hemoglobin as iron sources indicating that there is more than one heme transport system in those organisms (Nienaber et al. 2001; Wexler et al. 2001).

The energy for the transport of heme or siderophore-mediated iron transport across the outer membrane is provided by the TonB-ExbB-ExbD complex, where ExbBD somehow use the proton motive force to induce conformational changes in TonB. This energy coupling protein in turn interacts directly with the outer membrane TonB-dependent receptors (Postle and Larsen 2007). The precise mechanism involved is not completely understood. Moreover, it has been recently found that TonB is needed for transport of substrates in addition to iron (Schauer et al. 2008; Cornelis et al. 2009). In some organisms, there are many TonB-dependent receptors and only one TonB protein, while in other bacteria there are a few TonB-dependent receptors and more than one TonB protein (Schauer et al. 2008).

In *B. japonicum* 110, there are two clusters encoding TonB-ExbB-ExbD-like proteins, *bll7071-73* and *blr3906-08*. It is worth noting that *bll7071-bll7072* mutants are impaired in utilization of heme as an iron source, but can still use ferrichrome or ferric citrate as iron sources suggesting different functions for both TonB systems (Nienaber et al. 2001). A different situation is found in *R. leguminosarum,* where the disruption of the *tonB* homolog impedes the use of vicibactin as well as heme as iron sources (Wexler et al. 2001). In *S. meliloti* 1021 there is only one gene encoding a TonB-like protein and its function has not yet been reported.

3.2.2.2 HmuP

HmuP (*hemP*) is a small gene first identified incidentally many years ago in the sequencing of the heme transport gene cluster of *Yersinia entercolitica* (Stojiljkovic and Hantke 1992), and homologs have since been found in many Proteobacteria. However, clues into the function of *hmuP* have emerged only recently. A role for *hmuP* was identified in a screen for genes that affect expression of the outer membrane heme receptor gene *shmR* in *S. meliloti* (Amarelle et al. 2010). Expression of *shmR* in response to iron limitation is severely diminished in a *HmuP* mutant. The *shmR* gene is also controlled by RirA, but the H*muP* mutant is not defective in expression of other RirA-dependent genes. A regulatory function for H*muP* is interesting because it is found in bacteria known or suspected to have global regulators of iron homeostasis that also control heme transport. This suggests multiple levels of control of heme utilization gene expression.

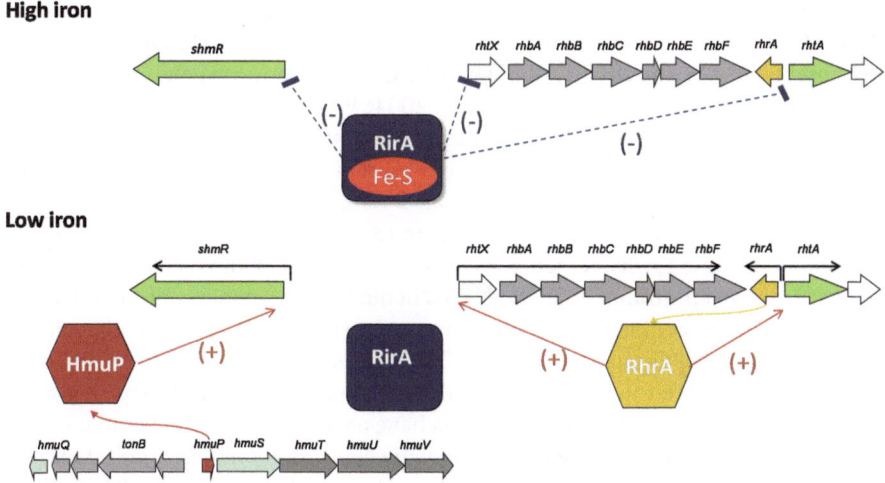

Fig. 3.4   Model of regulation of *shmR* and rhizobactin 1021 gene expression, under low and high iron conditions, in *S. meliloti* 1021. The RirA regulator may sense iron cellular status through a labile Fe atom of its Fe–S cluster, although this remains to be demonstrated experimentally. In the presence of iron, the RirA protein containing the Fe–S cluster represses expression of *shmR* gene as well as genes required for Rhizobactin 1021 biosynthesis (*rhbA*-F) and transport (*rhtA*). When cells are iron-deprived, the labile Fe is lost and the RirA regulator is not active as a repressor. Under this condition, the HmuP regulator activates the expression of *shmR* gene and RhrA activates the expression of genes required for rhizobactin 1021 biosynthesis and transport

Expression of the *S. meliloti shmR* gene under iron limitation likely involves both loss of function of the repressor RirA and the presence of an active HmuP protein (Fig. 3.4). By contrast, HmuP in *B. japonicum* functions as a co-activator with Irr to activate the HmuR operon (Escamilla-Hernandez and O'Brian 2012) (see discussion of Irr in section below). Although HmuP has not been studied in the numerous other bacteria that contain a H*muP* homolog, heme utilization genes are negatively regulated by Fur in those organisms. Thus, if HmuP plays a similar role in those bacteria as occurs in the rhizobia, thus heme utilization would likely require both derepression and activation. Further studies on the function of HmuP in the various organisms that contain different co-regulators will shed further light on this question.

3.2.2.3   Heme Degradation for Iron Release

Once exogenous heme is taken up by bacterial cells, the tetrapyrrole ring is cleaved by heme oxygenase in order to release the iron. The pioneering work on bacterial heme oxygenases was carried out with the proteins from *Corynebacterium diphtheria* (Schmitt 1997; Wilks and Schmitt 1998) and *Neisseriae meningitidis* (Zhu et al. 2000a, b; Ratliff et al. 2001; Schuller et al. 2001). These bacterial heme oxygenases have limited sequence similarity to each other and to eukaryotic heme oxygenases, but have similar structures overall (Schuller et al. 1999; Schuller et al. 2001; Hirotsu et al. 2004). They degrade heme to iron, CO, and biliverdin.

Subsequently, a structurally unrelated heme-degrading oxygenase was described in *Staphylococcus aureus* (Skaar et al. 2004; Wu et al. 2005), and is also found in a limited number of Gram-positive bacteria, including *Bacillus anthracis* (Skaar et al. 2006). Rather than biliverdin, the oxo-bilirubin chromophore staphylobilin is a product of *S. aureus* IsdG (Reniere et al. 2010).

Because numerous rhizobia can use heme as an iron source (Noya et al. 1997), they should be able to cleave heme and release iron. The photosynthetic bacterium *Bradyrhizobium* sp. strain ORS578 cleaves heme for the synthesis of the biliverdin chromophore, and contains a gene encoding a classical HmuO protein adjacent to a phytochrome gene (Giraud et al. 2000, 2002). It is not known whether this heme oxygenase is involved in iron utilization. *Rhodopseudomonas palustris, Agrobacterium tumefaciens,* and *R. leguminosarum* have a similar gene arrangement in their genomes. *B. japonicum* does not have an *hmuO* gene homolog, but it does have two genes, *hmuD* and *hmuQ*, encoding weak homologs to the *S. aureus* IsdG (Puri and O'Brian 2006). *hmuQ* is within the gene cluster encoding the heme transport system identified by Nienebar et al. (2001). hmuQ binds heme with high affinity and catalyzes the cleavage of heme to biliverdin (Puri and O'Brian, 2006). Homologs of *hmuQ* and *hmuD* were identified in many bacterial genera, and the recombinant homolog from *Brucella melitensis* has heme degradation activity as well. Thus, *hmuQ* and *hmuD* encode heme oxygenases, and the IsdG family of heme-degrading monooxygenases is not restricted to Gram-positive pathogenic bacteria.

Most bacterial mutants defective in heme oxygenases have no or mild phenotypes and retain the ability to use heme as an iron source. Similarly, a *Brucella abortus* mutant defective in the gene encoding BhuQ, an IsdG family protein, grows on heme, but displays elevated levels of siderophore, indicative of iron stress (Ojeda et al. 2012). Moreover, a *bhuQ* mutant has a severe growth phenotype in a siderophore synthesis background that is rescued by $FeCl_3$ but not heme. Those workers suggest that BhuQ may make the transcriptional regulator RirA (see below) more responsive to iron-derived heme. However, it should be considered that BhuQ has a function unrelated to heme degradation.

3.2.3 Ferric Dicitrate

Citric acid can complex ferric ion and it can be used for iron transport into bacterial cells, although the affinity of iron for citrate is not as high as it is for siderophores. Some bacteria can use citrate as a xenosiderophore (e.g. *E. coli, Mycobacterium smegmatis, Neisseria meningitidis,* and *Pseudomonas aeruginosa*) and, occasionally, it can be used as an endogenous siderophore. For instance *B. japonicum* strains 61A152 and 110 can use citrate as siderophore, but only *B. japonicum* 61A152 strain is able to secrete it (Guerinot et al. 1990). In *E. coli* and *P. aeruginosa,* iron uptake from ferric-dicitrate is inducible by environmental citrate, and this effect requires an extracytoplasmic function (ECF) sigma factor. The ferric-dicitrate uptake system requires the outer membrane receptor FecA, the periplasmic binding protein, and the ABC transporter system FecBCDE (Enz et al. 2000).

Although some rhizobia strains can use ferric citrate as an iron source, *B. japonicum* 110, *R. leguminosarum* 3841, and *S. meliloti* 1021 genomes have no homologs of genes encoding a FecABCDE citrate transport system. In *Streptomyces coelicolor*, exogenous citrate serves not only as an iron chelator, but also as an energy source, and thus ferric citrate is taken up for both purposes (Lensbouer et al. 2008). In this case, ferric citrate is taken up by a member of the CitMHS family of secondary transporters, a family found in numerous Gram-positive bacteria. No obvious homologs are found in sequenced rhizobial genomes.

3.2.4 Role of Iron Transport in Symbiosis

Symbiotic bacteroids have a large iron demand, but little is known about how iron gets into cells or the form of the metal that is transported. Mutations in genes encoding iron or heme transport systems described thus far do not have changed symbiotic phenotypes, indicating that such activities are not required for symbiotic function, or else they are redundant (Gill and Neilands 1989; Fabiano et al. 1995; Lynch et al. 2001; Nienaber et al. 2001; Wexler et al. 2001; Puri and O'Brian 2006; Amarelle et al. 2008; Cuiv et al. 2008; Small et al. 2009a, b). One exception is the *fegA* gene of *B. japonicum* strain 61A152, which is required for symbiosis (Benson et al. 2005). It was noted that accessibility to ferrichrome within a nodule is unlikely, so another function was suggested for this gene (Nienaber et al. 2001; Wexler et al. 2001; Benson et al. 2005). However, the corresponding gene in the USDA 122 derivative LO, *bll4920*, is not required for symbiosis, nor are the other four iron-responsive putative ferric siderophore receptor genes (Small et al. 2009a, b). Similarly, *S. meliloti* mutants defective in ferrichrome and ferrioxamine B uptake have no symbiotic phenotypes on alfalfa nodules (Cuiv et al. 2008).

The abundance of leghemoglobin in the infected plant cells of nodules makes it tempting to speculate that it is a potential iron source for bacteroids. Bacteroids are surrounded by a peribacteroid membrane to form symbiosomes, and thus the endosymbiont is physically separated from the plant globin. Heme transport across the peribacteroid membrane has not been demonstrated, and the heme transport systems employed by free-living rhizobia are not necessary for symbiosis (Nienaber et al. 2001; Wexler et al. 2001). However, leghemoglobin heme may be accessible to the bacteria during nodule senescence.

Isolated soybean symbiosomes can take up both ferric chelates and ferrous iron (Moreau et al. 1995; LeVier et al. 1996). Ferric chelate substrates are likely reduced to the ferrous form by a peribacteroid ferric reductase activity (LeVier et al. 1996), and ferrous iron can be taken up by bacteroids (Moreau et al. 1998). Soybeans express the divalent ion transporter GmDMT1 on the bacteroid membrane within nodules that may account for direct ferrous iron uptake by symbiososmes (Kaiser et al. 2003). Thus, it is possible that ferrous iron is the form taken up by the endosymbiont.

Ferrous iron is transported by FeoB in *E. coli* and other bacteria, and homologous systems have been identified in other organisms (Hantke 2003). A putative FeoB protein encoded by *blr6523* is present in the genome of *B. japonicum* 110, but its function remains unknown. There are no homologs of the Feo system encoded on the *S. meliloti* 1021 or *R. leguminosarum* 3481 genomes. The existence of a novel ferrous iron uptake system or other divalent metal transporter involved in ferrous iron uptake requires further investigation. As noted earlier, symbiotic bacteroids may receive an adequate supply of iron from the plant host, and may not normally be iron limited. In that case, a high affinity bacterial iron transport system would appear to be unnecessary.

3.3 Regulation of Iron Homeostasis in the Rhizobia

Studies on the control of bacterial iron homeostasis have focused largely on Fur (ferric uptake regulator), a regulatory protein that responds to cellular iron. Perhaps the biggest surprise in elucidating iron metabolism in the rhizobia is that genetic regulation differs greatly from the bacterial paradigm established by *E. coli* and other model systems. Although the reductionist approach has been enormously successful, this aspect of rhizobial biology reminds us of the diversity of bacterial biology, and the need for restraint in extrapolating data from a few model systems to an entire kingdom of life. Moreover, there is substantial diversity within the rhizobia. We now know that rhizobial Fur homologs are manganese-responsive regulators, at least in some species, and that the job of iron perception and regulation is carried out by two other regulators, Irr and RirA. Furthermore, some species have only Irr, whereas others have both Irr and RirA.

3.3.1 The Fur/Mur Protein

3.3.1.1 Brief Overview of Bacterial Fur

In *E. coli* and many other bacteria, Fur represses genes involved in high affinity iron transport under iron repleted conditions, and which are derepressed when the metal is scarce. In addition, Fur is involved in numerous other facets of iron metabolism, and also in processes not obviously linked to iron, such as acid shock response (Hall and Foster 1996), synthetic pathways (Stojiljkovic et al. 1994), and the production of toxins and other virulence factors (Litwin and Calderwood 1993). Fur is the founding member of a family of regulators which also includes Zur (Gaballa and Helmann 1998; Patzer and Hantke 1998), PerR (Bsat et al. 1998),

Irr (Hamza et al. 1998; Qi et al. 1999), and Nur (Ahn et al. 2006). These proteins differ in function and have different DNA binding sites, but are all involved in metal-dependent control of gene expression.

Fur homologs are found in many bacterial genomes. Structural analysis of Fur and its DNA binding properties have been most extensively studied in *E. coli*, *P. aeruginosa* and *B. subtilis*, whereas analyses of *fur* mutants and the identification of genes under Fur control have also been studied in those bacteria and in several other organisms as well. The working model for Fur function posits that, when bound by ferrous (Fe^{2+}) iron, Fur binds its target DNA within the promoter of the regulated gene to repress transcription. However, when iron is limiting in the cell, Fur protein is unbound by iron and no longer binds DNA with high affinity, hence gene expression is derepressed.

Downregulation of Fur- and iron-responsive genes in *fur* mutants implicates positive control. In numerous cases, apparent positive control of Fur-dependent genes is an indirect effect of its repression of a small RNA that negatively regulates genes in an iron-dependent manner (Masse and Gottesman 2002; Wilderman et al. 2004; Mellin et al. 2007; Gaballa et al. 2008; Mellin et al. 2010). In *E. coli*, Fur relieves silencing of the ferritin gene by the histone-like protein H-NS by binding to multiple sites on the ferritin promoter to prevent H-NS binding (Nandal et al. 2009). Other studies indicate direct positive control by demonstrating Fur binding to target genes (Delany et al. 2004; Grifantini et al. 2004; Ernst et al. 2005; Danielli et al. 2006; Gao et al. 2008; Butcher et al. 2011; Yu and Genco 2012), although recruitment of RNA polymerase or increasing promoter strength as a consequence of Fur binding have not been demonstrated. In addition, the metal-free form of Fur from *Helicobacter pylori*can bind DNA in several promoters (Delany et al. 2001, 2003; Ernst et al. 2005).

3.3.1.2 The Fur Homolog Mur is a Manganese-Responsive Regulator in the Rhizobia and Other α-Proteobacteria

The *sitABCD* operon of *S. meliloti* was identified in a screen for mutants that could not grow in the presence of metal chelator, and was shown to be necessary for manganese acquisition (Platero et al. 2003). These genes encode an ABC system belonging to the metal transport family found in many bacteria (Claverys 2001). A *fur* gene homolog lies adjacent to the *sitABCD* genes, which led those authors to speculate that Fur may control *sitABCD*. Subsequently, three reports came out showing that this is indeed the case, but the regulatory metal is Mn^{2+}, not Fe^{2+} (Chao et al. 2004; Diaz-Mireles et al. 2004; Platero et al. 2004). Manganese-responsiveness of a *sitA* reporter fusion is lost and constitutively high in a *fur* mutant of *R. leguminosarum* or *S. meliloti*, and Fur binds to the *sitA* promoter in vitro (Chao et al. 2004; Diaz-Mireles et al. 2004; Platero et al. 2004, 2007). Furthermore, the *sitABCD* operon is not regulated by iron. Microarray analysis of an *S. meliloti fur* mutant identified 23 genes that are normally down regulated by Fur (Chao et al. 2004). Other than *sitABCD*, it is not known whether these genes are regulated by manganese, or whether they have manganese-related functions. Mur has also been described in α-Proteobacteria other than the rhizobia. *B. abortus* Mur controls the *mntH* gene (Anderson et al. 2009; Menscher et al. 2012), and it suppressors the manganese transporter *sitABCD* operon *Agrobacterium tumefaciens* (Kitphati et al. 2007).

The *B. japonicum* Mur homolog was initially identified based on its ability to complement an *E. coli fur* mutant (Hamza et al. 1999). Mur binds the *irr* gene promoter, and a *mur* mutant shows loss of iron-responsive expression of *irr* mRNA (Hamza et al. 2000; Friedman and O'Brian 2003). In addition, global expression of numerous iron-responsive genes is aberrant in a *mur* mutant (Hamza et al. 2000; Yang et al. 2006c). Mur controls iron-dependent expression of the *fegA* gene encoding the ferrichrome receptor in *B. japonicum* strain 61A152 (Benson et al. 2004), but not in strains 110 or 122 (Small et al. 2009b). Finally, *B. japonicum* Mur is activated by either Fe^{2+} or Mn^{2+} in vitro to bind DNA and repress transcription (Friedman and O'Brian 2003, 2004). However, more recent studies show unequivocally that *B. japonicum* Mur is a manganese-responsive regulator, and its role in iron metalloregulation is probably minor or indirect in most cases.

B. japonicum Mur has three known direct targets, *irr*, *mntH*, and *mnoP*, and those genes contain a conserved motif in their promoters that binds Mur (Friedman et al. 2006; Hohle and O'Brian 2009, 2010; Hohle et al. 2011). Expression of the *mntH* and *mnoP* genes is responsive to manganese, but not iron. The apparent responsiveness of the *irr* gene to iron is now known to be due to the combined activity of Mur repression and Irr anti-repression (Hohle and O'Brian 2010). Early experiments were routinely carried out with manganese as part of the nutrient supplementation, which results in Mur occupancy of the *irr* promoter. Under iron limitation, Irr also binds to the *irr* promoter at a site that overlaps with the Mur-binding site, thus preventing Mur binding. Thus, transcript is high under iron limitation, and also in a *mur* mutant. Most other iron-regulated genes that are affected in a *mur* mutant only show modest effects, and have not been shown to be direct targets of the regulator (Yang et al. 2006c).

It is now clear that the Fur homolog is not the primary iron-responsive transcriptional regulator in the rhizobia, and that this function is taken over by Irr, RirA, or both. Thus rhizobia, and apparently many other α-proteobacteria, control iron homeostasis by a mechanism quite different from the paradigm established in *E. coli* and other model organisms.

3.3.1.3 Characterization of the Mur-binding Cis-Acting Regulatory Element

Central to the model of Fur function in *E. coli* and other model organisms is the so-called Fur box, a DNA binding element for Fur that contains similarity to a 19-bp, AT-rich palindromic consensus sequence (Fig. 3.4). Sequence similarity to a Fur box consensus within promoter regions of genes is taken as ab initio evidence for regulation by Fur. The binding site of Fur has been interpreted as two 9 bp inverted repeats, as three shorter hexameric repeats in a head-to-head-to-tail orientation, and as two 7-1-7 inverted repeat motifs (Escolar et al. 1998; Baichoo and Helmann 2002)

B. japonicum Mur was originally identified based on its ability to complement an *E. colifur* gene mutant, thus its ability to bind to a Fur box consensus is not surprising (Hamza et al. 1999; Friedman and O'Brian 2003). Similarly, the *E. coli* Fur-regulated *bfd* gene is regulated by *R. leguminosarum* Fur/Mur in an iron-dependent manner in *E. coli* (Diaz-Mireles et al. 2004). However, these organisms do not contain an *E. coli*-like Fur box sequence. Characterization of the promoters of the Mur-regulated genes *irr, mntH,* and *mnoP* in *B. japonicum* identified the binding site as three imperfect direct repeat hexamers that are all required for normal occupancy by Fur. This site is dissimilar to the *E. coli* Fur box consensus (Friedman and O'Brian 2003; Hohle and O'Brian 2009, 2010). Although *B. japonicum* Mur binds both this element and Fur box DNA with high affinity, *E. coli* Fur does not bind to the *irr* Mur-binding site. Mur binds to the promoter as one or two dimers, and this binding is sufficient to inhibit transcription in vitro in a metal-dependent manner (Friedman and O'Brian 2003).

The *R. leguminosarum* and *S. meliloti sitA* promoters have two and one Mur-binding sites, respectively, with similar sequences (Diaz-Mireles et al. 2004; 2005; Platero et al. 2007). These cis-acting elements are described as palindromic, and although they are somewhat similar to the Fur consensus sequence, Mur binds the Fur box with lower affinity (Wexler et al. 2003; Bellini and Hemmings 2006; Platero et al. 2007). *R. leguminosarum* Mur binds each *sitA* binding site as one or two dimers (Bellini and Hemmings 2006), but *S. meliloti* Mur occupies its cognate element as a single dimer (Platero et al. 2007). The basis for these differences is unknown, but the *S. meliloti* study used much less protein in the in vitro binding analysis than the *R. leguminosarum* Mur work. Mutation of a single hexamer within the *B. japonicum irr* or *mntH* Mur-binding site can result in occupancy of only one dimer (Friedman and O'Brian 2003; Hohle and O'Brian 2009), suggesting that differences in the DNA targets may account for differential occupancy, as has been described for *E. coli* Fur (Escolar et al. 1999). A *B. japonicum* Mur mutant was identified that can only occupy the *irr* promoter as a single dimer (Friedman and O'Brian 2004), but the residues that were altered are conserved in the *S. meliloti* and *R. leguminosarum* proteins. Despite differences between the *B. japonicum* Mur-binding sites and those of *S. meliloti* and *R. leguminosarum*, computational analysis of known and putative Mur-binding sites indicates a core of common residues in all of the *cis*-acting elements (Rodionov et al. 2006). A comparison of the five known Mur-binding sites yields a similar consensus (Fig. 3.4).

3.3.1.4 Metal-Binding Properties of Mur

Mur regulators function by binding the cognate regulatory metal directly, which confers DNA-binding activity on the protein (Friedman and O'Brian 2004; Bellini and Hemmings 2006; Platero et al. 2007). A regulatory Fe^{2+}-binding site (site 1) and a structural Zn^{2+}-binding site (site 2) implicated from the recent crystal structure of Fur from *P. aeruginosa* comprise amino acids highly conserved in many Fur family proteins, including those in the rhizobia. *B. japonicum* Mur mutants containing substitutions in site 1 or site 2 bound DNA with high affinity and repressed transcription in vitro in a metal-dependent manner. Interestingly, only a single dimer of site 2 mutant occupied the *irr* promoter, whereas the wild-type and site 1 mutant displayed one or two dimers occupancy. Both mutants were able to repress transcription from the *irr* promoter in vitro. Furthermore, both DNA binding and transcriptional repression were strictly metal-dependent. It appears that the putative functions for metal-binding sites deduced from the structure of *P. aeruginosa* Fur cannot be extrapolated to other bacterial Fur proteins as a whole.

3.3.1.5 Analyses of Mutants Suggests Novel Aspects of Manganese Uptake and Function

It is generally assumed that manganese is an essential nutrient, but recent studies on bacterial manganese transport mutants cast some doubt on whether this is generally true, and have shown differences between rhizobia and other organisms. MntH and MntABCD (SitABCD) are the two most widely represented Mn^{2+} transport systems in the eubacterial kingdom. MntH is the major high affinity Mn^{2+} transporter in numerous bacteria (Makui et al. 2000; Que and Helmann 2000; Domenech et al. 2002; Anderson et al. 2009; Hohle and O'Brian 2009), and the *mntH* gene is expressed under manganese limitation to scavenge available metal. Surprisingly, *mntH* mutants have no or mild growth phenotypes in numerous bacteria under non-stress conditions (Makui et al. 2000; Que and Helmann 2000; Domenech et al. 2002). Under normal growth, *E. coli* cells take up and contain little manganese, and manganese-dependent superoxide dismutase is not correctly metallated (Anjem et al. 2009). The lack of a manganese requirement suggests that manganese-dependent processes are not essential under those conditions, that other metals can substitute for manganese in manganese-containing proteins, or that these proteins are rendered non-essential due to compensatory activities.

B. japonicum has a single functional *mntH* gene, and no obvious *mntABC* gene homologs are present in the genome (Hohle and O'Brian 2009). A *B. japonicum mntH* mutant is almost completely defective in high affinity Mn^{2+} uptake activity. Moreover, the *mntH* strain has a severe growth phenotype under normal growth conditions (Hohle and O'Brian 2009), suggesting a greater reliance on manganese compared to *E. coli* and perhaps other organisms as well. Manganese as an essential nutrient may be a general feature of the Proteobacteria because *mntH* or *mntA* (*sitA*) mutants of *B. abortus* (Anderson et al. 2009) or *Sinorhizobium meliloti* (Platero et al. 2003; Davies and Walker 2007), respectively, also have growth phenotypes, and also a defect in manganese-dependent superoxide dismutase activity.

Nutritional metals such as manganese are available as the divalent cation in aerobic environments, and are thus soluble. Whereas cytoplasmic (inner) membrane transporters of free metal ions are well characterized in bacteria (Patzer and Hantke 1998; Kehres et al. 2000; Tottey et al. 2001; Degen and Eitinger 2002), translocation across the outer membrane in Gram-negative bacteria into the periplasm has not been described until recently. In principle, outer membrane pores with no or low selectivity should readily accommodate the diffusion of these small, soluble nutrients that are needed only in low quantities (Nikaido 2003; Silhavy et al. 2010). However, bacteria occupy niches in which the metal is often scarce, and *B. japonicum* cells can readily take up Mn^{2+} available in the low nanomolar range (Hohle and O'Brian 2009). Thus, simple diffusion across the outer membrane down such a shallow gradient via a non-selective pore is not likely to be sufficient to satisfy the nutritional needs of the cell. In support of this idea, MnoP was identified in *B. japonicum* based on its co-regulation with the inner membrane transporter gene *mntH* (Hohle et al. 2011). MnoP is an outer membrane protein expressed specifically under manganese limitation. MnoP acts as a channel to facilitate the tranlocation of Mn^{2+}, but not Co^{2+} or Cu^{2+}, into reconstituted proteoliposomes. An *mnoP* mutant is defective in high affinity Mn^{2+} transport into cells, and has a severe growth phenotype under manganese limitation. This suggests that the outer membrane is a barrier to divalent metal ions in Gram-negative bacteria that requires a selective channel to meet the nutritional needs of the cell.

B. japonicum, *S. meliloti,* and *B.rucella abortus* all have a single manganese-dependent superoxide dismutase that is expressed under unstressed growth (Santos et al. 1999; Davies and Walker 2007; Anderson et al. 2009; Hohle and O'Brian 2012), presumably to detoxify superoxide arising from aerobic respiration. The activities of these enzymes are all diminished in a manganese transport mutant. However, a *B. japonicum mntH* mutant has a much more severe growth phenotype than does a mutant defective in the superoxide dismutase gene *sodM* (Hohle and O'Brian 2012), implying that the defective activity cannot completely explain the *mntH* phenotype.

The growth phenotype of a *B. japonicum mntH* mutant is partially rescued by replacement of glycerol with pyruvate as the carbon source (Hohle and O'Brian 2012). This raises the possibility that glycerol utilization has a manganese-dependent step that is bypassed with pyruvate. Pyruvate kinase is a glycolytic enzyme that synthesizes pyruvate from phosphoenolpyruvate, and is required for utilization of glycerol, but not pyruvate, as an energy source. Animal pyruvate kinases are Mg^{2+}-dependent enzymes, but can use various divalent metals in vitro, including Mn^{2+}. *B. japonicum* has a single pyruvate kinase, PykM, and activity of that enzyme is deficient in a *mntH* strain. Moreover, purified PykM is activated by Mn^{2+} but not by other divalent metals (Hohle and O'Brian 2012). The *E. coli* pyruvate kinase PykF was activated by Mn^{2+} or Mg^{2+}, but only maintains allosteric control by fructose 1,6 bisphosphate in the presence of Mg^{2+} (Hohle and O'Brian 2012).

Pyruvate shuttles into numerous biosynthetic and energy-generating pathways, placing pyruvate kinase, hence manganese, at a crucial metabolic intersection. In this light, it is not surprising that *B. japonicum* has not only a high affinity inner membrane transporter for Mn^{2+}, but also the specific outer membrane channel

MnoP for uptake of the metal (Hohle et al. 2011). Unlike an *mntH* strain, a *pykM sodM* double mutant has only a minor growth phenotype with pyruvate as a carbon source, implying additional manganese-dependent processes that are yet to be determined.

3.3.2 The Irr Protein

3.3.2.1 Overview of Irr

Irr is prevalent in the α subdivision of the proteobacterial phylum. Among the sequenced genomes, it is ubiquitous in the order *Rhizobiales* and *Rhodobacteriales*, and found in some *Rhodospirillales* as well (reviewed in Rodionov et al. 2006). It is also present in the marine bacterium *Pelagibacter ubique*, which is in the order SAR11, but is not present in its obligate intracellular relatives *Rickettsia*, *Wolbachia*, and *Ehrlichia*. Interestingly, an Irr homolog is also found in *Acidothiobacillus ferrooxidans*, a γ-Proteobacterium that lives in acidic environments and is exposed to iron predominantly in the ferrous form. Microarray analysis shows that the vast majority of *B. japonicum* genes that are strongly regulated by iron are under the control of Irr (Yang et al. 2006b). Thus, Irr is the major iron regulator in that bacterium and probably in other *Bradyrhizobiaceae*, with Fur/Mur having a lesser or no role. Many α-proteobacteria within the *Rhizobiaceae* contain RirA in addition to Irr (Todd et al. 2002, 2005Yeoman et al. 2004; Chao et al. 2005; Viguier et al. 2005; Battisti et al. 2007; Ngok-Ngam et al. 2009; Hibbing and Fuqua 2011; Ojeda et al. 2012). The *rirA* gene is known or predicted to be controlled by Irr in numerous bacteria (Rodionov et al. 2006; Todd et al. 2006; Ojeda et al. 2012), and thus Irr may also affect the RirA regulon. Collectively, Irr and RirA appear to usurp the role of Fur in organisms that contain these novel regulators.

The *irr* gene was initially identified in *B. japonicum* in a screen for loss of control of heme biosynthesis by iron (Hamza et al. 1998), and it has been most extensively characterized in that organism. Heme is the end product of a biosynthetic pathway, culminating with the insertion of iron into a protoporphyrin ring to produce protoheme. Irr coordinates the heme biosynthetic pathway with iron availability to prevent the accumulation of toxic porphyrin precursors under iron limitation (Hamza et al. 1998). Loss of function of the *irr* gene is sufficient to uncouple the pathway from iron-dependent control, as discerned by the accumulation of protoporphyrin. This accumulation is due to derepression of *hemB* and probably *hemA* (Hamza et al. 1998; Yang et al. 2006b). Similarly, an *irr* mutant of *R. leguminosarum* has a fluorescent colony phenotype and is deregulated for the *hemA* gene (Wexler et al. 2003; Todd et al. 2006), and a *B.rucella abortus irr* mutant accumulates protoporphyrin (Martinez et al. 2005).

The Irr protein belongs to the Fur family of metalloregulators that includes Fur, PerR, Zur, Nur, and Mur (Bsat et al. 1998; Gaballa and Helmann 1998; Hamza et al. 1998; Patzer and Hantke 1998; Chao et al. 2004; Diaz-Mireles et al. 2004; Platero et al. 2004; Ahn et al. 2006). However, Irr behaves differently from these and other regulatory proteins in fundamentally different ways, and allows novel control of iron metabolism.

3.3.2.2 Irr is a Global Regulator of Iron Homeostasis

Although Irr was initially described in the context of heme biosynthesis, it is now clear that Irr is a global regulator of iron homeostasis and metabolism (Rudolph et al. 2006; Todd et al. 2006; Yang et al. 2006b). Transcriptome analysis of *B. japonicum* shows that Irr has a large regulon, and that most genes strongly controlled by iron at the mRNA level are also regulated by Irr (Yang et al. 2006b). Numerous Irr-regulated genes contain a *cis*-acting DNA element called an iron control element (ICE) within their promoters. The ICE motif was originally identified between the divergently transcribed genes *hmuR* and *hmuT* from *B. japonicum*, and shown to be necessary for activation of those genes under iron limitation (Nienaber et al. 2001; Rudolph et al. 2006b). Irr has been subsequently shown to bind numerous ICE-containing promoters in vitro and occupy those promoters in vivo (Yang et al. 2006b; Sangwan et al. 2008; Small et al. 2006b; Hohle and O'Brian 2010; Singleton et al. 2010; Escamilla-Hernandez and O'Brian 2012). Furthermore, bioinformatic analyses identified ICE-like motifs upstream of many open reading frames in *B. japonicum* and other α-Proteobacteria (Rodionov et al. 2006; Rudolph et al. 2006b).

Transcriptome analysis in *B. japonicum* also identified many Irr-regulated genes that apparently lack an upstream ICE motif (Yang et al. 2006b), suggesting either recognition of cis-elements dissimilar to ICE or indirect control by Irr. In support of the former, Irr binds promoter DNA lacking an ICE motif in *Brucella abortus* (Martinez et al. 2006, Anderson et al. 2011) and *Bartonella quintana* (Battisti et al. 2007; Parrow et al. 2009), although ICE-like motifs are predicted in numerous other genes in those organisms (Rodionov et al. 2006).

3.3.2.3 Mechanisms of Positive and Negative Control of Target Genes by Irr

Irr is both a positive and negative regulator of gene expression, and the mechanisms of control are understood to varying extents (Fig. 3.5). As expected, numerous negatively controlled genes are known or predicted to encode proteins that contain iron or are involved in the biosynthesis of heme or iron-sulfur clusters. Numerous positively controlled genes are involved in iron transport or iron-independent proteins that have an iron-dependent isozyme. It should be noted, however, that many Irr-dependent genes identified in transcriptome analysis encode proteins of unknown function (Yang et al. 2006b).

Negative control has been studied in two *B. japonicum* genes (Rudolph et al. 2006b, Sangwan et al. 2008), one encoding bacterioferritin (*bfr*), and the other,

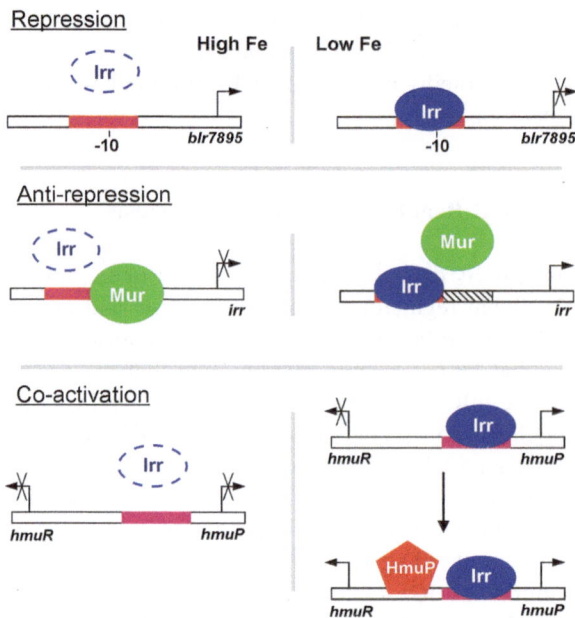

Fig. 3.5 Irr-mediated positive and negative control of iron-responsive genes. Cells grown under iron-replete conditions lack Irr activity due to degradation in *B. japonicum* and probably *B. abortus,* and by inactivation in *R. leguminosarum.* The model shown is based on *B. japonicum,* where the three types of control have been demonstrated. The shaded area of the DNA shows the Irr binding site, and the hashed area is the Mur-binding site. The bent arrow denotes the transcription start site. Although HmuP has been shown to occupy the *hmuR-hmuP* promoter in vivo, direct interaction with DNA or other proteins has not been demonstrated

blr7895, is a gene of unknown function, but a mutant in the homologous gene in *Agrobacterium tumefaciens* is sensitive to hydrogen peroxide stress (Ruangkiattikul et al. 2012). These genes are maximally expressed in the presence of iron in wild-type cells, and are derepressed in an *irr* mutant (Rudolph et al. 2006b; Sangwan et al. 2008). Correspondingly, Irr occupies the *bfr* and *blr7895* promoters under iron limitation (Sangwan et al. 2008), implicating a repressor role for the protein. Both gene promoters contain an ICE motif that overlaps the transcription site and are near or within the 10 regions (Rudolph et al. 2006a, b), and Irr binds those promoters with very high affinity in vitro (Sangwan et al. 2008). Moreover, in vitro transcription initiated from the *blr7895* promoter is inhibited by Irr (Sangwan et al. 2008). Collectively, the findings support a model of repression, whereby Irr occupancy of the target promoter is sufficient to repress expression, probably by occluding the promoter from RNA polymerase.

A case of anti-repression as a means of positive control by Irr has been described in *B. japonicum* (Hohle and O'Brian 2010) (Fig. 3.5). *Irr* mRNA levels are modestly controlled by iron, with maximal transcript found under iron limitation. The *irr* gene is occupied by Mur under manganese- and iron-replete conditions to repress transcription. However, when iron is limited, Irr accumulates and binds its own promoter, and Mur is unbound to it regardless of the manganese status. The Mur and Irr binding sites overlap on the *irr* promoter, and Irr occupancy prevents Mur binding in vivo and in vitro (Hohle and O'Brian 2010). Moreover, Mur-dependent transcription from the *irr* promoter in vitro is relieved in the presence of Irr. Irr is not necessary for high *irr* mRNA expression in a *mur* mutant, and thus the collective evidence shows that Irr is an anti-repressor rather than an activator. This conclusion cannot be directly extrapolated to other genes positively controlled by Irr, as most of them do not appear to be regulated by Mur (Yang et al. 2006b, c; Hohle et al. 2011). Nevertheless, Irr may be an anti-repressor of other negative regulators that are yet to be elucidated.

Many gene transcripts are downregulated in an *irr* mutant compared to the parent strain in *B. japonicum* (Yang et al. 2006b), and positive control by Irr has also been implicated in *B. abortus* (Anderson et al. 2011) and *Agrobacterium tumefaciens* (Hibbing and Fuqua 2011). Studies of positively controlled genes have focused primarily on those involved in the transport of iron or heme into cells (Nienaber et al. 2001; Martinez et al. 2006; Rudolph et al. 2006b, Small et al. 2006b; Anderson et al. 2011; Small and O'Brian 2011; Escamilla-Hernandez and O'Brian 2012). In *B. japonicum*, the promoters of these genes are occupied by Irr under iron limitation in vivo (Small et al. 2006b; Escamilla-Hernandez and O'Brian 2012), and thus the control exerted by Irr is direct. In addition, promoters each contain an ICE motif that is distal from the transcription start site or $-10/-35$ region (Rudolph et al. 2006b; Small et al. 2006b), providing at least an inference that Irr functions by activation rather than anti-repression of those genes (Barnard et al. 2004).

Although the mechanism of activation by Irr is unknown, recent work identified HmuP as a regulator of Irr-dependent expression of the *hmuR* operon encoding proteins needed for heme utilization as a nutritional iron source (Fig. 3.5). The *hmuP* gene is found divergently from the *hmuR* operon in the context of the *hmuPTUV* operon (Escamilla-Hernandez and O'Brian 2012). A single ICE motif is found between the two divergent operons, and iron-dependent control of both operons is impaired when that motif is mutated (Nienaber et al. 2001). Moreover, Irr binds the *hmuR-hmuP* promoter region in vivo (Escamilla-Hernandez and O'Brian 2012). An *hmuP* deletion abrogates activation of the *hmuR* operon, but substantial iron-dependent control of the divergent *hmuPTUV* operon remains. HmuP binds the *hmuR-hmuP* promoter, but cannot induce the *hmuR* operon in the absence of Irr. These findings implicate HmuR as a co-activator of Irr-dependent activation of the *hmuR* operon.

3.3.2.4 Irr is Controlled Post-translationally by a Heme-dependent Mechanism

Irr functions under iron limitation, which makes it distinct from other Fur family proteins because it functions in the absence of the regulatory metal, whereas other members require direct metal binding for activity in most cases.

The Irr protein accumulates in cells under iron limitation, with very low levels in iron-replete cells in *B. japonicum* and *B.rucella abortus* (Hamza et al. 1998; Martinez et al. 2005; Anderson et al. 2011), and the mechanism controlling Irr levels have been worked out in some detail in *B. japonicum*. In that organism, Irr is a conditionally stable protein that degrades in cells exposed to iron (Qi et al. 1999). *B. japonicum* Irr contains a heme-regulatory motif (HRM) near the N-terminus that binds heme and is necessary for rapid degradation. Accordingly, Irr is stabilized in a heme-deficient background or by mutagenesis of Cys-29 within the HRM. Since the discovery of heme-dependent degradation of Irr, numerous other eukaryotic proteins have been identified that degrade in response to heme by binding to HRM motifs (Jeong et al. 2004; Ishikawa et al. 2005; Zenke-Kawasaki et al. 2007; Hu et al. 2008; Yang et al. 2008).

B. japonicum Irr fused to glutathione S transferase (GST) confers iron-dependent instability on GST, but a GST fusion containing only the N-terminal 36 amino acids of Irr, which includes the HRM, is stable (Yang et al. 2005). This means that the HRM is necessary but not sufficient for rapid degradation of Irr. In addition, a C29A mutant within the HRM eventually degrades after long-term exposure to iron, but the wild-type protein is completely stable in a heme-deficient strain. This implicates an additional heme-dependent degradation process independent of the HRM. In vitro and in vivo studies identified an instability domain that includes three consecutive histidines at positions 117-119, with His-117 and His-119 being invariant residues in Irr proteins. Mutation of the HXH domain results in a very stable protein independent of iron. This HXH domain is part of a heme-binding region distinct from the HRM. Raman and EPR spectroscopy confirm two heme binding sites, one with a cysteine coordinated axial ligand and the second with six coordinate His/His ligation (Ishikawa et al. 2011).

The activity of *R. leguminosarum* Irr is also heme-dependent, but heme does not trigger protein degradation (Singleton et al. 2010). Instead, heme binding to the conserved HXH motif decreases its affinity for target DNA. Examination of the Irr homologs reveals that those within the *Bradyrhizobiaceae, Xanthobacteraceae, Methylobacteriaceae, Rhodospirillaceae and Beijerinckiaceae* have the Cys-Pro sequence and an HRM-like domain, and the HXH motif corresponding to His-117 and His-119 of *B. japonicum* Irr are completely conserved in all of the homologs. Many Irr homologs, including that of *R.hizobium leguminosarum*, lack the HRM. This raises the question of whether Irr degradation as described for *B. japonicum* occurs in other rhizobia. Singleton et al. (Singleton et al. 2010) suggested that heme-dependent degradation of Irr may correlate with an HRM, and those that lack an HRM may result in heme-dependent inactivation rather than degradation as described for the *R. leguminosarum* Irr. This is unlikely in light of the fact that Irr accumulates only under iron limitation in *B. abortus*, which does not contain an HRM (Martinez et al. 2005; Anderson et al. 2011). In addition, the *B. japonicum* Irr derivative lacking an HRM degrades, but the rate is much slower than is found for the wild-type protein (Yang et al. 2005). *B. japonicum* has a lower affinity ferric heme binding site (Qi and O'Brian 2002) that could possibly serve a similar function as the HRM, albeit less efficiently. By analogy, Irr homologs lacking an HRM may have a compensatory mechanism that allows turnover.

3.3.2.5 Regulated Degradation of *B. japonicum* Irr Requires both Redox States of Heme

Whereas the HRM binds specifically to ferric (Fe^{3+}) heme, the histidine-rich domain binds ferrous (Fe^{2+}) heme (Yang et al. 2005). An Irr mutant in whom the three histidines are replaced by alanines is stable in vivo under iron-replete conditions (Yang et al. 2005). Irr decay follows first order kinetics (Qi et al. 1999), indicating a single mechanism for degradation. Hence the two hemes likely participate in a single degradation process rather than independent processes that occur at different rates. These findings implicate a role for the redox activity of heme in Irr degradation, and further evidence suggests that this activity leads to protein oxidation (Yang et al. 2006a). *B. japonicum* Irr degrades in response to cellular oxidative stress by a mechanism that involves heme and iron (Yang et al. 2006a). Furthermore, Irr degradation is strictly O_2-dependent in vivo (Yang et al.2006a). Irr oxidation was demonstrated in vitro, requiring heme, O_2 and a reductant. An Irr truncation that does not bind ferrous heme in vitro does not degrade in vivo. Thus, it was proposed that reactive oxygen species participate in Irr degradation not only as part of an oxidative stress response (see below), but also in normal degradation in response to iron. Protein oxidation can result in hydrolysis of peptide bonds (Berlett and Stadtman 1997) and thus, in principle, oxidation of Irr could be sufficient for degradation. However, in vivo degradation of Irr is rapid, whereas carbonylation in vitro is slow. It is probable that oxidized Irr is recognized by cellular proteases as a damaged protein that is subsequently degraded. A candidate protease has not been described thus far. Degradation of Bach1 requires the ubiquitin ligase HOIL-1, which interacts with the heme-bound form of the respective protein (Ishikawa et al. 2005; Zenke-Kawasaki et al. 2007). The ubiquitin-tagged protein is then degraded. Arginyl transferase is tagged by N-end rule ubiquitin ligases in yeast and mouse for heme-dependent degradation (Hu et al. 2008).

Redox-dependent ligand switching, although not associated with protein degradation, occurs with the transcriptional regulator CooA from *Rhodospirillum rubrum* (Roberts et al. 2004) and the redox sensor EcDos from *E. coli* (Kurokawa et al. 2004). IRP2 binds ferric heme at the HRM and ferrous heme at a conserved histidine residue (Jeong et al. 2004; Ishikawa et al. 2005), and protein oxidation is reported to be heme-dependent (Yamanaka et al. 2003). However, subsequent work strongly indicates iron-dependent degradation of IRP2 by a heme-independent mechanism that does not involve the HRM (Salahudeen et al. 2009; Vashisht et al. 2009). In those studies, IRP was shown to be degraded by an E3 ubiquitin ligase complex that includes the iron and oxygen-responsive protein FBXL5. FBXL5 contains a hemerythrin-like domain at its N-terminus that confers instability under iron or oxygen deprivation, thus the protein accumulates when those compounds are high.

3.3.2.6 *B. japonicum* Irr Interacts with Heme Locally at the Site of Heme Synthesis

A fundamental problem with heme as a signaling molecule is that it is reactive and lipophilic. Heme can catalyze the formation of reactive oxygen species, and binds non-specifically to lipids, proteins, and other macromolecules. Thus, a regulatory free heme pool is unlikely. The discovery of new and novel roles for heme as a regulatory molecule in eukaryotes and prokaryotes begs for reconciliation between these functions and the cytotoxicity of heme. This problem has been partially resolved for the Irr protein from *B. japonicum*. Ferrochelatase catalyzes the insertion of iron into protoporphyrin to form heme in the final step of the heme biosynthetic pathway. Irr interacts directly with ferrochelatase and responds to iron via the status of heme and protoporphyrin localized at the site of heme synthesis (Qi and O'Brian 2002) (Fig. 3.5). Competition of the wild-type ferrochelatase with a catalytically inactive one inhibits iron-dependent degradation of Irr even though the cell is not heme-defective. This means that Irr does not respond to a free heme pool, but rather to heme locally where it is synthesized. The dissociation binding constant (K_d) of heme for Irr is about 1 nM, which is less than one free heme molecule per cell. Irr may represent the simplest type of heme signaling mechanism, because there is no obvious need for a factor to chaperone heme from the site of synthesis to its target.

The interaction of Irr with ferrochelatase is affected by the immediate heme precursor protoporphyrin. The porphyrin-bound enzyme does not bind to Irr, which is the state of ferrochelatase when iron is limiting, and allows Irr to be active and affect the genes under its control. Thus, Irr is affected by heme and by its substrates so that heme synthesis does not exceed iron availability. In the presence of iron, ferrochelatase inactivates Irr, followed by Irr degradation to derepress the pathway. Irr is present but inactive in cells that express a catalytically inactive ferrochelatase, but active in a *hemH* deletion strain (Qi and O'Brian 2002). It is possible that inactivation of Irr allows loss of function that is faster than its degradation. Indeed, the *hemB* mRNA is elevated by iron more rapidly than Irr degrades (Chauhan et al. 1997; Qi et al. 1999).

3.3.2.7 Iron Homeostasis is Controlled by the Status of Heme via Irr

Irr interacts directly with the heme biosynthesis enzyme ferrochelatase, resulting in degradation under iron-replete conditions, or accumulation of active protein under iron limitation (Qi and O'Brian 2002). Thus, the discovery that Irr is a global regulator of iron-regulated genes indicates that iron homeostasis is controlled by the status of heme. Indeed, a heme-deficient strain of *B. japonicum* cannot maintain normal iron homeostasis. Control of Irr-regulated genes is aberrant in a heme-defective *B. japonicum* mutant, resulting in iron-replete cells behaving as if they are iron-limited (Yang et al. 2006b). The heme mutant has abnormally high cellular iron content, probably because iron transport genes are constitutively activated due to persistence of Irr in that strain. Accordingly, under iron limitation an *irr* mutant behaves as if it were iron replete, even though cellular iron levels are lower than that found in the wild-type (Yang et al. 2006b).

Most bacteria studied to date sense and respond to iron directly to regulate gene expression. That is, iron binds directly to a regulatory protein to modulate its activity. Iron binding to Fur confers DNA-binding activity on the protein, as also occurs for the DtxR regulator from *Corynebacterium diphtheriae* and its homolog, the IdeR protein from *Mycobacterium tuberculosis* (Escolar et al. 1999; Pohl et al. 1999a, b). However, *B. japonicum*, and perhaps other α-proteobacteria, do not sense iron directly, but rather sense and respond to an iron-dependent process, namely the biosynthesis of heme. Is there an advantage to this type of control? Approximately one-half of the total iron in iron-limited *B. japonicum* cells is found in heme (unpublished observations). Since heme biosynthesis places such a high energy demand on the cell, this synthesis may serve as a sensitive indicator of the overall iron status. Also, many iron-dependent processes such as electron transport, tricarboxylic acid cycle, and detoxification are associated with aerobic metabolism, which also requires heme. Therefore, it may allow a better coordination of cellular events.

3.3.2.8 Oxidative Stress Promotes Degradation of the *B. japonicum* Irr Protein

Bacteria have multiple defense strategies against oxidative stress, including the direct detoxification of ROS by catalase, peroxidases, and superoxide dismutase. Oxidative stress responses require the activation of regulatory proteins and the induction of genes under their control. In many bacteria, the transcriptional regulator OxyR (Christman et al. 1989; Tao et al. 1989) senses hydrogen peroxide (Zheng et al. 1998) and induces numerous genes whose products are involved in peroxide defense (Tartaglia et al. 1989; Altuvia et al. 1994), redox balance (Prieto-Alamo et al. 2000; Ritz et al. 2000) and other factors (Altuvia et al. 1997; Zheng et al. 1999). In *B. subtilis*, PerR is the major peroxide regulator and represses a large PerR regulon (Herbig and Helmann 2001). The OhrR family of antioxidant regulators is responsible for organic hydroperoxide resistance (Mongkolsuk et al. 1998). *B. japonicum* contains an OxyR homolog, but it may function differently in that organism than in other systems (Panek and O'Brian 2004). Evidence points to Irr as an oxidative stress response regulator. Irr degrades in response to H_2O_2 produced endogenously in a catalase-deficient (*katG*) strain, or to H_2O_2 applied exogenously to culture media (Yang et al. 2006b). A *B. abortus irr* mutant displays elevated catalase activity and resistance to killing by H_2O_2 (Martinez et al. 2006). The Irr deficiency causes derepression of *hemB* in *B. japonicum* and elevated heme in *B. abortus*. Catalases and peroxidases are heme proteins that detoxify H_2O_2 and peroxides, respectively, and elevated *hemB* may contribute to the synthesis of those enzymes. Other examples of elevated expression of heme biosynthesis genes are noted. The *Bacillus subtilis* PerR protein mediates the induction of the *hemAXCDBL* operon encoding enzymes for the early steps of heme synthesis (Chen et al. 1995; Mongkolsuk and Helmann 2002). In *E. coli* and *Salmonella*, the *hemH* gene encoding the heme biosynthetic enzyme ferrochelatase is induced in response to H_2O_2 in an OxyR-dependent manner (Zheng et al. 2001; Elgrably-Weiss et al. 2002). However, it has not been established in *B. japonicum* or any other bacterium that synthesis of catalase or peroxidase substantially increases the overall heme demand in the cell, and thus the physiological relevance of elevated heme synthesis genes is uncertain.

3.3.2.9 Irr is Involved in the Coordination of Iron and Manganese Homeostasis

Recent work indicates that the metabolism of manganese and iron are interrelated in prokaryotes and eukaryotes, and some mechanisms and rationale for the relationship is emerging. Manganese protects cells against oxidative stress, and iron has pro-oxidative properties, which provides the basis for at least some aspects of the relationship between the two metals. Manganese can substitute for iron in numerous *E. coli* enzymes, rendering those proteins less sensitive to hydrogen peroxide damage (Sobota and Imlay 2011; Anjem and Imlay 2012). This may be a general protective mechanism for mononuclear iron enzymes, which do not require the redox activity of iron for their activity. Manganese can also substitute for iron in a class Ib ribonucleotide reductase under conditions where iron is limited and manganese is available (Martin and Imlay 2011).

In *B. japonicum*, severe manganese limitation created by growth of an Mn^{2+} transport mutant in manganese limited media requires more iron for growth compared to wild-type cells (Puri et al. 2010). Correspondingly, manganese limitation results in a cellular iron deficiency. Irr is a positive regulator of iron transport, and Irr levels are attenuated under manganese limitation in wild-type cells, resulting in reduced promoter occupancy of target genes, and altered iron-dependent gene expression. Thus, manganese control of the iron status is mediate through Irr.

Irr levels remain high regardless of manganese availability in a heme-deficient mutant, indicating that manganese normally affects heme-dependent degradation of Irr. Manganese alters the secondary structure of Irr in vitro, and inhibits binding of heme to the protein. Because heme is required for Irr degradation, the data suggest that manganese stabilizes Irr by inhibiting heme binding. Thus, under manganese limitation, Irr is destabilized under low iron conditions by lowering the threshold of heme that can trigger Irr degradation.

The Mn^{2+} transporter MntH is required for growth of iron-deficient cells in *E. coli*, which has been interpreted in terms of manganese substituting for iron in activating mononuclear enzymes (Anjem et al. 2009). In that case, iron deficiency creates a need for manganese import to compensate for the lack of iron. The *B. japonicum* work also shows a requirement for *mntH* in low iron of *B. japonicum* if manganese is also deficient, but the reasons for this appear to be different from that described in *E. coli*. In *B. japonicum*, manganese limitation causes iron limitation which can be partially rescued by increasing iron availability (Puri et al. 2010). These observations do not suggest substitution of one metal for the other, but rather a mechanism for decreasing the iron content when manganese is limiting.

It is plausible that manganese limitation renders cells more vulnerable to oxidative stress, and that attenuating the cellular iron content limits the effects of iron on oxidative stress.

The *irr* gene is transcriptionally regulated by both manganese and iron through Mur and Irr respectively, as described above (Hohle and O'Brian 2010) (Fig.3.5). Irr relieves Mur-dependent repression under iron limitation when manganese is present, and those are the conditions of maximal Irr activity. Irr binds manganese to be fully functional, and thus the anti-repression mechanism ensures maximal transcription. However, the rationale for multiple levels of control is not entirely clear in light of the fact that primary control of the *irr* gene is post-translational. One possibility is that changes in *irr* mRNA under low iron conditions increases the rate of response, but does not appreciably affect the steady-state level. Alternatively, the transcriptional control may contain an evolutionary vestige. An ancestral form of the *fur* gene may have been autoregulated in a negative manner, as has been shown in *E. coli* (Escolar et al. 1999), and *irr* arose from gene duplication of *fur*. As both Irr and Mur changed function, control by Mur was maintained but an additional anti-repressor function evolved to maintain basal mRNA level, which is necessary for post-transcriptional control.

3.3.3 The RirA Protein

The realization that Fur is not a global regulator of iron transport in *Rhizobium leguminosarum* led Johnston and colleagues to search for mutants that showed deregulation of genes that are transcriptionally controlled by iron, leading to the discovery of RirA (Todd et al. 2002). They found that numerous iron-regulated genes are constitutively high in a *rirA* mutant, indicating negative control by RirA. Similarly, RirA controls expression of the rhizobactin siderophore synthesis operon in *S. meliloti* (Viguier et al. 2005), and microarray analysis reveals a large RirA regulon that includes numerous genes involved in iron transport, energy metabolism, and exopolysaccharide production (Chao et al. 2005). Thus, some systems transcriptionally repressed by Fur in *E. coli* are negatively controlled by RirA in *S. meliloti* and *R. leguminosarum*. Although most genes downregulated by iron limitation in a microarray study are independent of RirA in *S. meliloti* (Chao et al. 2005), an analysis of the *R. leguminosarum* proteome reveals 17 proteins that are diminished in an *rirA* strain (Todd et al. 2005).

A *S. meliloti rirA* mutant has a growth phenotype in iron-replete media that is partially restored under iron limitation (Chao et al. 2005). Similarly, the growth phenotype of an *Agrobacterium tumefaciens rirA* strain is relieved by a second mutation in *irr*, a gene that functions under iron limitation (Hibbing and Fuqua

2011). The growth deficiency of the *rirA* mutants in the presence of iron may be caused by the accumulation of iron to toxic levels as a result of elevated iron transport activity. Consistent with this, the *S. meliloti rirA* strain is more sensitive to H_2O_2 in the presence of iron, presumably due to the generation of reactive oxygen species via the Fenton reaction (Chao et al. 2005). In addition, a *rirA* mutant of *A. tumefaciens* constitutively expresses iron uptake genes, and is more sensitive than the wild-type to the iron-activated antibiotic streptonigrin and to hydrogen peroxide (Ngok-Ngam et al. 2009).

As described above, the rhizobia can use heme as an iron source, which requires both its transport into cells, and cleavage of the macrocycle to release iron. A *rirA* mutant is unable to grow in the presence of heme (Chao et al. 2005). The *hemPSTU* operon is constitutively expressed in the mutant, indicating that lack of growth is due to heme toxicity rather than iron deficiency (Chao et al. 2005). This conclusion is supported by the inability to rescue the growth phenotype of the *rirA* mutant in the presence of heme by addition of inorganic iron.

RirA has not yet been studied in vitro, and thus its exact mechanism of function has not been elucidated. RirA belongs to the Rrf2 family of putative transcriptional regulators, a family that is not well characterized as a whole. The best described member is IscR, a transcriptional regulator that represses iron–sulfur cluster assembly gene operons, and activates the *suf* operon (Schwartz et al. 2001; Giel et al. 2006; Yeo et al. 2006). IscR is an iron–sulfur protein, and the three cysteines are conserved in RirA (Todd et al. 2002). IscR is an active protein both in the absence and presence of the iron–sulfur cluster. Repression of the *iscRSUA* operon requires the iron–sulfur center, but the demetallated protein activates *sufABCDFE*. The lability of the iron–sulfur cluster links its assembly to the control of O_2-regulated genes (Giel et al. 2006). Most relevant to the current discussion is that IscR responds to the iron and sulfur status (Outten et al. 2004; Gyaneshwar et al. 2005). Johnston and colleagues speculate that RirA may respond to iron through the status of an iron–sulfur cluster (Todd et al. 2006). The generality of IscR function to Rrf2 family members as a whole is not yet known. It will be important to determine whether RirA is an iron–sulfur protein, and whether the status of that moiety is sensitive to the cellular iron status.

By examining the promoters of several *R. leguminosarum* genes deregulated in an *rirA* mutant, a consensus RirA-dependent cis-acting regulatory element was identified (Yeoman et al. 2004). Deletion or mutation of these elements results in loss of iron-responsiveness. Furthermore, a bioinformatic search identified RirA-responsive elements upstream of numerous genes in *R. leguminosarum* and *S. meliloti* known to be regulated by RirA (Rodionov et al. 2006). Several iron-regulated *R. leguminosarum* genes are constitutive when introduced into *Paracoccus denitrificans* on a plasmid, but iron-responsiveness is restored when the *rirA* gene is also introduced (Yeoman et al. 2004). Collectively, it is very likely that RirA binds directly to these promoters to repress gene expression in the presence of iron.

RirA is prevalent in the *Rhizobiaceae*, which includes *Brucella, Bartonella,* and *Agrobacterium,* as well as *Sinorhizobium* and *Rhizobium,* but homologs are not confined to this taxonomic group. Proteins with high similarity (at least 60 % similarity) to *R. leguminosarum* RirA are present in *Nitrobacter* (Bradyrhizobiaceae), *Labrenzia* (Rhodobacterales), and in species within the more distant γ- and β-Proteobacteria, and Firmicutes. The prevalence of RirA within the Rhizobiaceae might suggest that it is the primary iron-responsive regulator within this group. However, studies of several species do not confirm this idea according to their authors. As described above, synthesis of the *B. abortus* siderophores brucebactin and 2,3-dihydroxybenzoic acid is controlled by Irr (Martinez et al. 2005, 2006). However, a putative RirA binding site upstream of the *dhbCEBAD* operon encoding 2,3 dihydroxybenzoic acid synthesis proteins was identified in a bioinformatic study (Rodionov et al. 2006). Thus, a role for RirA in control of *B. abortus* siderophore gene expression is possible. In *B. quintana*, genes encoding heme-binding proteins (Battisti et al. 2007) and a heme utilization locus (Parrow et al. 2009) are regulated by Irr rather than by RirA. In the latter study, control of gene expression was assessed in wild type cells that overexpressed *fur, irr,* or *rirA* in trans. In *A. tumefaciens*, genes that are upregulated under iron limitation appear to be controlled by both Irr and RirA, whereas the *hemA* gene, which is down-regulated under low iron conditions, is regulated only by Irr (Hibbing and Fuqua 2011). Finally, numerous *rirA* homologs in the *Rhizobiaceae* contain a putative Irr binding site 5' of their coding regions (Rodionov et al. 2006), and control of *R. leguminosarum rirA* by Irr has been demonstrated directly (Todd et al. 2006).

The genes encoding the regulators HmuP and RhrA are derepressed in an *rirA* mutant (Chao et al. 2005). HmuP is in turn necessary to positively control the heme receptor gene *shmR* (Amarelle et al. 2010). The *shmR* gene has a putative RirA binding site (Rodionov et al. 2006), suggesting that *shmR* expression requires direct negative and positive control by RirA and HmuP, respectively (Fig. 3.4). Similarly, the rhizobactin 1021 synthesis operon *rhtXrhbABCDEF* is positively controlled by RhrA, and a putative RirA binding site upstream of rhtX suggests direct control by RirA, in addition to indirect control through the *rhrA* gene (Lynch et al. 2001; Cuiv et al. 2004) (Fig. 3.4). DNA binding studies of RirA are sorely needed to directly address those ideas.

Acknowledgments Research from the authors' laboratories was supported by a grant from PEDECIBA-Uruguay to E.F., NIH grants GM067966 and GM099667 to M.R.O'B.

References

Ahn BE, Cha J, Lee EJ, Han AR, Thompson CJ, Roe JH (2006) Nur, a nickel-responsive regulator of the Fur family, regulates superoxide dismutases and nickel transport in *Streptomyces coelicolor*. Mol Microbiol 59:1848–1858

Altuvia S, Almiron M, Huisman G, Kolter R, Storz G (1994) The *dps* promoter is activated by OxyR during growth and by IHF and sigma S in stationary phase. Mol Microbiol 13:265–272

Altuvia S, Weinstein-Fischer D, Zhang A, Postow L, Storz G (1997) A small, stable RNA induced by oxidative stress: role as a pleiotropic regulator and antimutator. Cell 90:43–53

Amarelle V, Koziol U, Rosconi F, Noya F, O'Brian MR, Fabiano E (2010) A new small regulatory protein, HmuP, modulates haemin acquisition in *Sinorhizobium meliloti*. Microbiology 156:1873–1882

Amarelle V, O'Brian MR, Fabiano E (2008) ShmR is essential for utilization of heme as a nutritional iron source in *Sinorhizobium meliloti*. Appl Environ Microbiol 74:6473–6475

Anderson ES, Paulley JT, Gaines JM, Valderas MW, Martin DW, Menscher E, Brown TD, Burns CS, Roop RM 2nd (2009) The manganese transporter MntH is a critical virulence determinant for *Brucella abortus* 2308 in experimentally infected mice. Infect Immun 77:3466–3474

Anderson ES, Paulley JT, Martinson DA, Gaines JM, Steele KH, Roop RM II (2011) The iron-responsive regulator Irr Is required for wild-type expression of the gene encoding the heme transporter BhuA in *Brucella abortus* 2308. J Bacteriol 193:5359–5364

Anjem A, Imlay JA (2012) Mononuclear iron enzymes are primary targets of hydrogen peroxide stress. J Biol Chem 287:15544–15556

Anjem A, Varghese S, Imlay JA (2009) Manganese import is a key element of the OxyR response to hydrogen peroxide in *Escherichia coli*. Mol Microbiol 72:844–858

Baginsky C, Brito B, Imperial J, Palacios JM, Ruiz-Argueso T (2002) Diversity and evolution of hydrogenase systems in rhizobia. Appl Environ Microbiol 68:4915–4924

Baichoo N, Helmann JD (2002) Recognition of DNA by Fur: a reinterpretation of the Fur box consensus sequence. J Bacteriol 184:5826–5832

Barnard A, Wolfe A, Busby S (2004) Regulation at complex bacterial promoters: how bacteria use different promoter organizations to produce different regulatory outcomes. Curr Opin Microbiol 7:102–108

Barsomian GD, Urzainqui A, Lohman K, Walker GC (1992) *Rhizobium meliloti* mutants unable to synthesize anthranilate display a novel symbiotic phenotype. J Bacteriol 174:4416–4426

Battisti JM, Smitherman LS, Sappington KN, Parrow NL, Raghavan R, Minnick MF (2007) Transcriptional regulation of the heme binding protein gene family of *Bartonella quintana* is accomplished by a novel promoter element and iron response regulator. Infect Immun 75:4373–4385

Battistoni F, Platero R, Duran R, Cervenansky C, Battistoni J, Arias A, Fabiano E (2002a) Identification of an iron-regulated, hemin-binding outer membrane protein in *Sinorhizobium meliloti*. Appl Environ Microbiol 68:5877–5881

Battistoni F, Platero R, Noya F, Arias A, Fabiano E (2002b) Intracellular Fe content influences nodulation competitiveness of *Sinorhizobium meliloti* strains as inocula of alfalfa. Soil Biol Biochem 34:593–597

Bellini P, Hemmings AM (2006) In vitro characterization of a bacterial manganese uptake regulator of the Fur superfamily. Biochemistry 45:2686–2698

Benson HP, Boncompagni E, Guerinot ML (2005) An iron uptake operon required for proper nodule development in the *Bradyrhizobium japonicum*-soybean symbiosis. Mol Plant Microbe Interact 18:950–959

Benson HP, LeVier K, Guerinot ML (2004) A dominant-negative *fur* mutation in *Bradyrhizobium japonicum*. J Bacteriol 186:1409–1414

Berlett BS, Stadtman ER (1997) Protein oxidation in aging, disease and oxidative stress. J Biol Chem 272:20313–20316

Bsat N, Herbig A, Casillas-Martinez L, Setlow P, Helmann JD (1998) *Bacillus subtilis* contains multiple Fur homologs: identification of the iron uptake (Fur) and peroxide regulon (PerR) repressors. Mol Microbiol 29:189–198

Burkhardt R, Braun V (1987) Nucleotide sequence of the *fhuC* and *fhuD* genes involved in iron (III) hydroxamate transport: domains in FhuC homologous to ATP-binding proteins. Mol Gen Genet 209:49–55

Butcher BG, Bronstein PA, Myers CR, Stodghill PV, Bolton JJ, Markel EJ, Filiatrault MJ, Swingle B, Gaballa A, Helmann JD, Schneider DJ, Cartinhour SW (2011) Characterization of the *fur* regulon in *Pseudomonas syringae* pv. tomato DC3000. J Bacteriol 193:4598–4611

Carlton TM, Sullivan JT, Stuart GS, Hutt K, Lamont IL, Ronson CW (2007) Ferrichrome utilization in a mesorhizobial population: microevolution of a three-locus system. Environ Microbiol 9:2923–2932

Carson KC, Meyer J-M, Dilworth MJ (2000) Hydroxamate siderophores of root nodule bacteria. Soil Biol Biochem 32:11–21

Carter RA, Worsley PS, Sawers G, Challis GL, Dilworth MJ, Carson KC, Lawrence JA, Wexler M, Johnston AW, Yeoman KH (2002) The vbs genes that direct synthesis of the siderophore vicibactin in *Rhizobium leguminosarum*: their expression in other genera requires ECF sigma factor RpoI. Mol Microbiol 44:1153–1166

Cescau S, Cwerman H, Letoffe S, Delepelaire P, Wandersman C, Biville F (2007) Heme acquisition by hemophores. Biometals 20:603–613

Challis GL (2005) A widely distributed bacterial pathway for siderophore biosynthesis independent of nonribosomal peptide synthetases. ChemBioChem 6:601–611

Chao TC, Becker A, Buhrmester J, Puhler A, Weidner S (2004) The *Sinorhizobium meliloti fur* gene regulates, with dependence on Mn(II), transcription of the *sitABCD* operon, encoding a metal-type transporter. J Bacteriol 186:3609–3620

Chao TC, Buhrmester J, Hansmeier N, Puhler A, Weidner S (2005) Role of the regulatory gene *rirA* in the transcriptional response of *Sinorhizobium meliloti* to iron limitation. Appl Environ Microbiol 71:5969–5982

Chauhan S, Titus DE, O'Brian MR (1997) Metals control activity and expression of the heme biosynthesis enzyme δ-aminolevulinic acid dehydratase in *Bradyrhizobium japonicum*. J Bacteriol 179:5516–5520

Chen L, Keramati L, Helmann JD (1995) Coordinate regulation of *Bacillus subtilis* peroxide stress genes by hydrogen peroxide and metal ions. Proc Natl Acad Sci U S A 92:8190–8194

Chen WM, Moulin L, Bontemps C, Vandamme P, Bena G, Boivin-Masson C (2003) Legume symbiotic nitrogen fixation by beta-proteobacteria is widespread in nature. J Bacteriol 185:7266–7272

Christman MF, Storz G, Ames BN (1989) OxyR, a positive regulator of hydrogen peroxide-inducible genes in *Escherichia coli* and *Salmonella typhimurium*, is homologous to a family of bacterial regulatory proteins. Proc Natl Acad Sci U S A 86:3484–3488

Claverys JP (2001) A new family of high-affinity ABC manganese and zinc permeases. Res Microbiol 152:231–243

Cornelis P, Matthijs S, Van Oeffelen L (2009) Iron uptake regulation in *Pseudomonas aeruginosa*. Biometals 22:15–22

Coulton JW, Mason P, Allatt DD (1987) *fhuC* and *fhuD* genes for iron (III)-ferrichrome transport into *Escherichia coli* K-12. J Bacteriol 169:3844–3849

Cuiv PO, Clarke P, Lynch D, O'Connell M (2004) Identification of *rhtX* and *fptX*, novel genes encoding proteins that show homology and function in the utilization of the siderophores rhizobactin 1021 by *Sinorhizobium meliloti* and pyochelin by *Pseudomonas aeruginosa*, respectively. J Bacteriol 186:2996–3005

Cuiv PO, Keogh D, Clarke P, O'Connell M (2008) The *hmuUV* genes of *Sinorhizobium meliloti* 2011 encode the permease and ATPase components of an ABC transport system for the utilization of both haem and the hydroxamate siderophores, ferrichrome and ferrioxamine B. Mol Microbiol 70:1261–1273

Dancis A, Klausner RD, Hinnebusch AG, Barriocanal JG (1990) Genetic evidence that ferric reductase is required for iron uptake in *Saccharomyces cerevisiae*. Mol Cell Biol 10:2294–2301

Danielli A, Roncarati D, Delany I, Chiarini V, Rappuoli R, Scarlato V (2006) In vivo dissection of the Helicobacter pylori Fur regulatory circuit by genome-wide location analysis. J Bacteriol 188:4654–4662

Davies BW, Walker GC (2007) Disruption of *sitA* compromises *Sinorhizobium meliloti* for manganese uptake required for protection against oxidative stress. J Bacteriol 189:2101–2109

Degen O, Eitinger T (2002) Substrate specificity of nickel/cobalt permeases: insights from mutants altered in transmembrane domains I and II. J Bacteriol 184:3569–3577

Delany I, Rappuoli R, Scarlato V (2004) Fur functions as an activator and as a repressor of putative virulence genes in *Neisseria meningitidis*. Mol Microbiol 52:1081–1090

Delany I, Spohn G, Rappuoli R, Scarlato V (2001) The Fur repressor controls transcription of iron-activated and -repressed genes in *Helicobacter pylori*. Mol Microbiol 42:1297–1309

Delany I, Spohn G, Rappuoli R, Scarlato V (2003) An anti-repression Fur operator upstream of the promoter is required for iron-mediated transcriptional autoregulation in *Helicobacter pylori*. Mol Microbiol 50:1329–1338

Diaz-Mireles E, Wexler M, Sawers G, Bellini D, Todd JD, Johnston AW (2004) The Fur-like protein Mur of *Rhizobium leguminosarum* is a Mn^{2+}-responsive transcriptional regulator. Microbiology 150:1447–1456

Diaz-Mireles E, Wexler M, Todd JD, Bellini D, Johnston AW, Sawers RG (2005) The manganese-responsive repressor Mur of *Rhizobium leguminosarum* is a member of the Fur-superfamily that recognizes an unusual operator sequence. Microbiology 151:4071–4078

Dilworth MJ (1980) [74] Leghemoglobins. In: Methods in enzymology, Vol 69. Academic Press, Salt Lake City, pp 812–823

Dilworth MJ, Carson KC, Giles RGF, Byrne LT, Glenn AR (1998) *Rhizobium leguminosarum* bv. viciae produces a novel cyclic trihydroxamate siderophore, vicibactin. Microbiology 144:781–791

Domenech P, Pym AS, Cellier M, Barry CE 3rd, Cole ST (2002) Inactivation of the *Mycobacterium tuberculosis* Nramp orthologue (*mntH*) does not affect virulence in a mouse model of tuberculosis. FEMS Microbiol Lett 207:81–86

Donadio S, Monciardini P, Sosio M (2007) Polyketide synthases and nonribosomal peptide synthetases: the emerging view from bacterial genomics. Nat Prod Rep 24:1073–1109

Downie JA, Walker SA (1999) Plant responses to nodulation factors. Curr Opin Plant Biol 2:483–489

Elgrably-Weiss M, Park S, Schlosser-Silverman E, Rosenshine I, Imlay J, Altuvia S (2002) A *Salmonella enterica* serovar *typhimuriumhemA* mutant is highly susceptible to oxidative DNA damage. J Bacteriol 184:3774–3784

Enz S, Mahren S, Stroeher UH, Braun V (2000) Surface signaling in ferric citrate transport gene induction: interaction of the FecA, FecR, and FecI regulatory proteins. J Bacteriol 182:637–646

Ernst FD, Bereswill S, Waidner B, Stoof J, Mäder U, Kusters JG, Kuipers EJ, Kist M, van Vliet AHM, Homuth G (2005a) Transcriptional profiling of *Helicobacter pylori* Fur- and iron-regulated gene expression. Microbiology 151:533–546

Ernst FD, Homuth G, Stoof J, Mäder U, Waidner B, Kuipers EJ, Kist M, Kusters JG, Bereswill S, van Vliet AHM (2005b) Iron-responsive regulation of the *Helicobacter pylori* iron-cofactored superoxide dismutase SodB Is mediated by Fur. J Bacteriol 187:3687–3692

Escamilla-Hernandez R, O'Brian MR (2012) HmuP is a co-activator of Irr-dependent expression of heme utilization genes in *Bradyrhizobium japonicum*. J Bacteriol 194:3137–3143

Escolar L, Perez-Martin J, de Lorenzo V (1998) Binding of the fur (ferric uptake regulator) repressor of *Escherichia coli* to arrays of the GATAAT sequence. J Mol Biol 283:537–547

Escolar L, Perez-Martin J, de Lorenzo V (1999) Opening the iron box: transcriptional metalloregulation by the Fur protein. J Bacteriol 181:6223–6229

Ettema TJG, Andersson SGE (2009) The α-proteobacteria: the Darwin finches of the bacterial world. Biol Lett 5:429–432

Expert D, Gill PRJ (1992) Iron: a modulator in bacterial virulence and symbiotic nitrogen-fixation. CRC Press, Boca Raton

Fabiano E, Gill PR Jr, Noya F, Bagnasco P, De La Fuente L, Arias A (1995) Siderophore-mediated iron acquisition mutants in *Rhizobium meliloti* 242 and its effect on the nodulation kinetic of alfalfa nodules. Symbiosis 19:197–211

Finking R, Marahiel MA (2004) Biosynthesis of nonribosomal peptides1. Annu Rev Microbiol 58:453–488

Friedman DB, Stauff DL, Pishchany G, Whitwell CW, Torres VJ, Skaar EP (2006) *Staphylococcus aureus* redirects central metabolism to increase iron availability. PLoS Pathog 2:e87

Friedman YE, O'Brian MR (2003) A novel DNA-binding site for the ferric uptake regulator (Fur) protein from *Bradyrhizobium japonicum*. J Biol Chem 278:38395–38401

Friedman YE, O'Brian MR (2004) The ferric uptake regulator (Fur) protein from *Bradyrhizobium japonicum* is an iron-responsive transcriptional repressor in vitro. J Biol Chem 279:32100–32105

Gaballa A, Antelmann H, Aguilar C, Khakh SK, Song K-B, Smaldone GT, Helmann JD (2008) The *Bacillus subtilis* iron-sparing response is mediated by a Fur-regulated small RNA and three small, basic proteins. Proc Natl Acad Sci U S A 105:11927–11932

Gaballa A, Helmann JD (1998) Identification of a zinc-specific metalloregulatory protein, Zur, controlling zinc transport operons in *Bacillus subtilis*. J Bacteriol 180:5815–5821

Gao H, Zhou D, Li Y, Guo Z, Han Y, Song Y, Zhai J, Du Z, Wang X, Lu J, Yang R (2008) The iron-responsive Fur regulon in *Yersinia pestis*. J Bacteriol 190:3063–3075

Genco CA, Dixon DW (2001) Emerging strategies in microbial haem capture. Mol Microbiol 39:1–11

Georgiadis MM, Komiya H, Chakrabarti P, Woo D, Kornuc JJ, Rees DC (1992) Crystallographic structure of the nitrogenase iron protein from *Azotobacter vinelandii*. Science 257:1653–1659

Gibson KE, Kobayashi H, Walker GC (2008) Molecular determinants of a symbiotic chronic infection. Annu Rev Genet 42:413–441

Giel JL, Rodionov D, Liu M, Blattner FR, Kiley PJ (2006) IscR-dependent gene expression links iron-sulphur cluster assembly to the control of O_2-regulated genes in *Escherichia coli*. Mol Microbiol 60:1058–1075

Gill PR Jr, Neilands JB (1989) Cloning a genomic region required for a high-affinity iron-uptake system in Rhizobium meliloti 1021. Mol Microbiol 3:1183–1189

Giraud E, Fardoux J, Fourrier N, Hannibal L, Genty B, Bouyer P, Dreyfus B, Vermeglio A (2002) Bacteriophytochrome controls photosystem synthesis in anoxygenic bacteria. Nature 417:202–205

Giraud E, Hannibal L, Fardoux J, Vermeglio A, Dreyfus B (2000) Effect of Bradyrhizobium photosynthesis on stem nodulation of *Aeschynomene sensitiva*. Proc Natl Acad Sci U S A 97:14795–14800

Grifantini R, Frigimelica E, Delany I, Bartolini E, Giovinazzi S, Balloni S, Agarwal S, Galli G, Genco C, Grandi G (2004) Characterization of a novel *Neisseria meningitidis* Fur and iron-regulated operon required for protection from oxidative stress: utility of DNA microarray in the assignment of the biological role of hypothetical genes. Mol Microbiol 54:962–979

Gruber N, Galloway JN (2008) An earth-system perspective of the global nitrogen cycle. Nature 451:293–296

Guerinot ML, Meidl EJ, Plessner O (1990) Citrate as a siderophore in *Bradyrhizobium japonicum*. J Bacteriol 172:3298–3303

Gyaneshwar P, Paliy O, McAuliffe J, Popham DL, Jordan MI, Kustu S (2005) Sulfur and nitrogen limitation in *Escherichia coli* K-12: specific homeostatic responses. J Bacteriol 187:1074–1090

Hall HK, Foster JW (1996) The role of fur in the acid tolerance response of *Salmonella typhimurium* is physiologically and genetically separable from its role in iron acquisition. J Bacteriol 178:5683–5691

Hamza I, Chauhan S, Hassett R, O'Brian MR (1998) The bacterial Irr protein is required for coordination of heme biosynthesis with iron availability. J Biol Chem 273:21669–21674

Hamza I, Hassett R, O'Brian MR (1999) Identification of a functional *fur* gene in *Bradyrhizobium japonicum*. J Bacteriol 181:5843–5846

Hamza I, Qi Z, King ND, O'Brian MR (2000) Fur-independent regulation of iron metabolism by Irr in *Bradyrhizobium japonicum*. Microbiol. 146:669–676

Hantke K (2003) Is the bacterial ferrous iron transporter FeoB a living fossil? Trends Microbiol 11:192–195

Hennecke H (1992) The role of respiration in symbiotic nitrogen fixation. Kluwer Academic Publishers, Dordrecht

Herbig AF, Helmann JD (2001) Roles of metal ions and hydrogen peroxide in modulating the interaction of the *Bacillus subtilis* PerR peroxide regulon repressor with operator DNA. Mol Microbiol 41:849–859

Herrada G, Puppo A, Moreau S, Day DA, Rigaud J (1993) How is leghemoglobin involved in peribacteroid membrane degradation during nodule senescence? FEBS Lett 326:33–38

Hibbing ME, Fuqua C (2011) Antiparallel and interlinked control of cellular iron levels by the Irr and RirA regulators of *Agrobacterium tumefaciens*. J Bacteriol 193:3461–3472

Hirotsu S, Chu GC, Unno M, Lee DS, Yoshida T, Park SY, Shiro Y, Ikeda-Saito M (2004) The crystal structures of the ferric and ferrous forms of the heme complex of HmuO, a heme oxygenase of *Corynebacterium diphtheriae*. J Biol Chem 279:11937–11947

Hohle TH, Franck WL, Stacey G, O'Brian MR (2011) Bacterial outer membrane channel for divalent metal ion acquisition. Proc Natl Acad Sci U S A 108:15390–15395

Hohle TH, O'Brian MR (2009) The *mntH* gene encodes the major Mn^{2+} transporter in *Bradyrhizobium japonicum* and is regulated by manganese via the Fur protein. Mol Microbiol 72:399–409

Hohle TH, O'Brian MR (2010) Transcriptional control of the *Bradyrhizobium japonicum irr* gene requires repression by Fur and antirepression by Irr. J Biol Chem 285:26074–26080

Hohle TH, O'Brian MR (2012) Manganese is required for oxidative metabolism in unstressed *Bradyrhizobium japonicum* cells. Mol Microbiol 84:766–777

Hu RG, Wang H, Xia Z, Varshavsky A (2008) The N-end rule pathway is a sensor of heme. Proc Natl Acad Sci U S A 105:76–81

Hu X, Boyer GL (1995) Isolation and characterization of the siderophore N-deoxyschizokinen from *Bacillus megaterium* ATCC 19213. Biometals 8(357):364

Ishikawa H, Kato M, Hori H, Ishimori K, Kirisako T, Tokunaga F, Iwai K (2005) Involvement of heme regulatory motif in heme-mediated ubiquitination and degradation of IRP2. Mol Cell 19:171–181

Ishikawa H, Nakagaki M, Bamba A, Uchida T, Hori H, O'Brian MR, Iwai K, Ishimori K (2011) Unusual heme binding in the bacterial iron response regulator protein: spectral characterization of heme binding to the heme regulatory motif. Biochemistry 50:1016–1022

Jeong J, Rouault TA, Levine RL (2004) Identification of a heme-sensing domain in iron regulatory protein 2. J Biol Chem 279:45450–45454

Kaiser BN, Moreau S, Castelli J, Thomson R, Lambert A, Bogliolo S, Puppo A, Day DA (2003) The soybean NRAMP homologue, GmDMT1, is a symbiotic divalent metal transporter capable of ferrous iron transport. Plant J 35:295–304

Kehres DG, Zaharik ML, Finlay BB, Maguire ME (2000) The NRAMP proteins of *Salmonella typhimurium* and *Escherichia coli* are selective manganese transporters involved in the response to reactive oxygen. Mol Microbiol 36:1085–1100

Kitphati W, Ngok-Ngam P, Suwanmaneerat S, Sukchawalit R, Mongkolsuk S (2007) *Agrobacterium tumefaciens fur* has important physiological roles in iron and manganese homeostasis, the oxidative stress response, and full virulence. Appl Environ Microbiol 73:4760–4768

Koster W, Braun V (1990) Iron (III) hydroxamate transport into *Escherichia coli*. Substrate binding to the periplasmic FhuD protein. J Biol Chem 265:21407–21410

Kunze B, Trowitzsch-Kienast W, Hofle G, Reichenbach H (1992) Nannochelins A, B and C, new iron-chelating compounds from Nannocystis exedens (myxobacteria). Production, isolation, physico-chemical and biological properties. J Antibiot (Tokyo) 45:147–150

Kurokawa H, Lee DS, Watanabe M, Sagami I, Mikami B, Raman CS, Shimizu T (2004) A redox-controlled molecular switch revealed by the crystal structure of a bacterial heme PAS sensor. J Biol Chem 279:20186–20193

Lankford CE (1973) Bacterial assimilation of iron. Crit Rev Microbiol 2:273–331

Lensbouer JJ, Patel A, Sirianni JP, Doyle RP (2008) Functional characterization and metal ion specificity of the metal-citrate complex transporter from *Streptomyces coelicolor*. J Bacteriol 190:5616–5623

Letoffe S, Delepelaire P, Wandersman C (2006) The housekeeping dipeptide permease is the *Escherichia coli* heme transporter and functions with two optional peptide binding proteins. Proc Natl Acad Sci U S A 103:12891–12896

Letoffe S, Delepelaire P, Wandersman C (2008) Functional differences between heme permeases: Serratia marcescens HemTUV permease exhibits a narrower substrate specificity (restricted to heme) than the *Escherichia coli* DppABCDF peptide-heme permease. J Bacteriol 190:1866–1870

LeVier K, Day DA, Guerinot ML (1996) Iron uptake by symbiosomes from soybean root nodules. Plant Physiol 111:893–900

Litwin CM, Calderwood SB (1993) Role of iron in regulation of virulence genes. Clin Microbiol Rev 6:137–149

Lynch D, O'Brien J, Welch T, Clarke P, Cuiv PO, Crosa JH, O'Connell M (2001) Genetic organization of the region encoding regulation, biosynthesis, and transport of rhizobactin 1021, a siderophore produced by *Sinorhizobium meliloti*. J Bacteriol 183:2576–2585

Makui H, Roig E, Cole ST, Helmann JD, Gros P, Cellier MF (2000) Identification of the *Escherichia coli* K-12 Nramp orthologue (MntH) as a selective divalent metal ion transporter. Mol Microbiol 35:1065–1078

Martin JE, Imlay JA (2011) The alternative aerobic ribonucleotide reductase of *Escherichia coli*, NrdEF, is a manganese-dependent enzyme that enables cell replication during periods of iron starvation. Mol Microbiol 80:319–334

Martinez M, Ugalde RA, Almiron M (2005) Dimeric *Brucella abortus* Irr protein controls its own expression and binds haem. Microbiology 151:3427–3433

Martinez M, Ugalde RA, Almiron M (2006) Irr regulates brucebactin and 2,3-dihydroxybenzoic acid biosynthesis, and is implicated in the oxidative stress resistance and intracellular survival of *Brucella abortus*. Microbiology 152:2591–2598

Masse E, Gottesman S (2002) A small RNA regulates the expression of genes involved in iron metabolism in *Escherichia coli*. Proc Natl Acad Sci U S A 99:4620–4625

Masson-Boivin C, Giraud E, Perret X, Batut J (2009) Establishing nitrogen-fixing symbiosis with legumes: how many rhizobium recipes? Trends Microbiol 17:458–466

Matzanke BF, Anemuller S, Schunemann V, Trautwein AX, Hantke K (2004) FhuF, part of a siderophore-reductase system. Biochemistry 43:1386–1392

McKie AT (2008) The role of Dcytb in iron metabolism: an update. Biochem Soc Trans 36:1239–1241

Mellin JR, Goswami S, Grogan S, Tjaden B, Genco CA (2007) A novel Fur- and iron-regulated small RNA, NrrF, is required for indirect Fur-mediated regulation of the sdhA and sdhC Genes in Neisseria meningitidis. J Bacteriol 189:3686–3694

Mellin JR, McClure R, Lopez D, Green O, Reinhard B, Genco C (2010) Role of Hfq in iron-dependent and -independent gene regulation in *Neisseria meningitidis*. Microbiology 156:2316–2326

Menscher EA, Caswell CC, Anderson ES, Roop RM (2012) Mur regulates the gene encoding the manganese transporter MntH in *Brucella abortus* 2308. J Bacteriol 194:561–566

Modi M, Shah KS, Modi VV (1985) Isolation and characterisation of catechol-like siderophore from cowpea Rhizobium RA-1. Arch Microbiol 141:156–158

Mongkolsuk S, Helmann JD (2002) Regulation of inducible peroxide stress responses. Mol Microbiol 45:9–15

Mongkolsuk S, Praituan W, Loprasert S, Fuangthong M, Chamnongpol S (1998) Identification and characterization of a new organic hydroperoxide resistance (ohr) gene with a novel pattern of oxidative stress regulation from *Xanthomonas campestris* pv. phaseoli. J Bacteriol 180:2636–2643

Moreau S, Day DA, Puppo A (1998) Ferrous iron is transported across the peribacteroid membrane of soybean nodules. Planta 207:83–87

Moreau S, Meyer JM, Puppo A (1995) Uptake of iron by symbiosomes and bacteroids from soybean nodules. FEBS Lett 361:225–228

Moulin L, Munive A, Dreyfus B, Boivin-Masson C (2001) Nodulation of legumes by members of the beta-subclass of Proteobacteria. Nature 411:948–950

Nandal A, Huggins CC, Woodhall MR, McHugh J, Rodriguez-Quinones F, Quail MA, Guest JR, Andrews SC (2009) Induction of the ferritin gene (*ftnA*) of *Escherichia coli* by Fe^{2+}-Fur is mediated by reversal of H-NS silencing and is RyhB independent. Mol Microbiol 75(3): 637–657

Neilands JB (1973) Microbial iron transport compounds (siderochromes). Elsevier, Amsterdam

Neilands JB (1981) Microbial iron compounds. Annu Rev Biochem 50:715–731

Neilands JB, Leong SA (1986) Siderophores in relation to plant growth and disease. Annu Rev Plant Physiol 37:187–208

Ngok-Ngam P, Ruangkiattikul N, Mahavihakanont A, Virgem SS, Sukchawalit R, Mongkolsuk S (2009) Roles of *Agrobacterium tumefaciens* RirA in iron regulation, oxidative stress response, and virulence. J Bacteriol 191:2083–2090

Nienaber A, Hennecke H, Fischer HM (2001) Discovery of a haem uptake system in the soil bacterium *Bradyrhizobium japonicum*. Mol Microbiol 41:787–800

Nikaido H (2003) Molecular basis of bacterial outer membrane permeability revisited. Microbiol Mol Biol Rev 67:593–656

Noya F, Arias A, Fabiano E (1997) Heme compounds as iron sources for nonpathogenic rhizobium bacteria. J Bacteriol 179:3076–3078

O'Brian MR, Thony-Meyer L (2002) Biochemistry, regulation and genomics of haem biosynthesis in prokaryotes. Adv Microb Physiol 46:257–318

O'Hara GW, Dilworth MJ, Boonkerd JN, Parkpian P (1988) Iron-deficiency specifically limits nodule development in peanut inoculated with *Bradyrhizobium sp*. New Phytol 108:51–57

Ojeda JF, Martinson D, Menscher E, Roop RM 2nd (2012) The *bhuQ* gene encodes a heme oxygenase that contributes to the ability of *Brucella abortus* 2308 to use heme as an iron source and is regulated by Irr. J Bacteriol (in press)

Oke V, Long SR (1999) Bacteroid formation in the *Rhizobium*-legume symbiosis. Curr Opin Microbiol 2:641–646

Okujo N, Sakakibara Y, Yoshida T, Yamamoto S (1994) Structure of acinetoferrin, a new citrate-based dihydroxamate siderophore from *Acinetobacter haemolyticus*. Biometals 7:170–176

Oldroyd GE, Downie JA (2008) Coordinating nodule morphogenesis with rhizobial infection in legumes. Annu Rev Plant Biol 59:519–546

Outten FW, Djaman O, Storz G (2004) A *suf* operon requirement for Fe-S cluster assembly during iron starvation in Escherichia coli. Mol Microbiol 52:861–872

Panek HR, O'Brian MR (2004) KatG is the primary detoxifier of hydrogen peroxide produced by aerobic metabolism in *Bradyrhizobium japonicum*. J Bacteriol 186:7874–7880

Parrow NL, Abbott J, Lockwood AR, Battisti JM, Minnick MF (2009) Function, regulation, and transcriptional organization of the hemin utilization locus of *Bartonella quintana*. Infect Immun 77:307–316

Patel HN, Chakraborty RN, Desai SB (1988) Isolation and partial characterization of phenolate siderophore from *Rhizobium leguminosarum* IARI 102. FEMS Microbiol Lett 56:131–134

Patzer SI, Hantke K (1998) The ZnuABC high-affinity zinc uptake system and its regulator Zur in *Escherichia coli*. Mol Microbiol 28:1199–1210

Persmark M, Pittman P, Buyer JS, Schwyn B, Gill PR Jr, Neilands JB (1993) Isolation and structure of rhizobactin 1021, a siderophore from the alfalfa symbiont *Rhizobium meliloti* 1021. J Am Chem Soc 115:3950–3956

Platero R, de Lorenzo V, Garat B, Fabiano E (2007) *Sinorhizobium meliloti fur*-like (Mur) protein binds a fur box-like sequence present in the *mntA* promoter in a manganese-responsive manner. Appl Environ Microbiol 73:4832–4838

Platero R, Peixoto L, O'Brian MR, Fabiano E (2004) Fur is involved in manganese-dependent regulation of mntA (sitA) expression in *Sinorhizobium meliloti*. Appl Environ Microbiol 70:4349–4355

Platero RA, Jaureguy M, Battistoni FJ, Fabiano ER (2003) Mutations in *sitB* and *sitD* genes affect manganese-growth requirements in *Sinorhizobium meliloti*. FEMS Microbiol Lett 218:65–70

Plessner O, Klapatch T, Guerinot ML (1993) Siderophore utilization by *Bradyrhizobium japonicum*. Appl Environ Microbiol 59:1688–1690

Pohl E, Holmes RK, Hol WG (1999a) Crystal structure of a cobalt-activated diphtheria toxin repressor-DNA complex reveals a metal-binding SH3-like domain. J Mol Biol 292:653–667

Pohl E, Holmes RK, Hol WG (1999b) Crystal structure of the iron-dependent regulator (IdeR) from *Mycobacterium tuberculosis* shows both metal binding sites fully occupied. J Mol Biol 285:1145–1156

Poole K (2004) Iron transport systems in pathogenic bacteria: pseudomonas. ASM Press, Washington

Postle K, Kadner RJ (2003) Touch and go: tying TonB to transport. Mol Microbiol 49:869–882

Postle K, Larsen R (2007) TonB-dependent energy transduction between outer and cytoplasmic membranes. Biometals 20:453–465

Prieto-Alamo MJ, Jurado J, Gallardo-Madueno R, Monje-Casas F, Holmgren A, Pueyo C (2000) Transcriptional regulation of glutaredoxin and thioredoxin pathways and related enzymes in response to oxidative stress. J Biol Chem 275:13398–13405

Puri S, Hohle TH, O'Brian MR (2010) Control of bacterial iron homeostasis by manganese. Proc Natl Acad Sci U S A 107:10691–10695

Puri S, O'Brian MR (2006) The *hmuQ* and *hmuD* genes from *Bradyrhizobium japonicum* encode heme-degrading enzymes. J Bacteriol 188:6476–6482

Qi Z, Hamza I, O'Brian MR (1999) Heme is an effector molecule for iron-dependent degradation of the bacterial iron response regulator (Irr) protein. Proc Natl Acad Sci U S A 96:13056–13061

Qi Z, O'Brian MR (2002) Interaction between the bacterial iron response regulator and ferrochelatase mediates genetic control of heme biosynthesis. Mol Cell 9:155–162

Que Q, Helmann JD (2000) Manganese homeostasis in *Bacillus subtilis* is regulated by MntR, a bifunctional regulator related to the diphtheria toxin repressor family of proteins. Mol Microbiol 35:1454–1468

Ratledge C, Dover LG (2000) Iron metabolism in pathogenic bacteria. Annu Rev Microbiol 54:881–941

Ratliff M, Zhu W, Deshmukh R, Wilks A, Stojiljkovic I (2001) Homologues of neisserial heme oxygenase in gram-negative bacteria: degradation of heme by the product of the *pigA* gene of *Pseudomonas aeruginosa*. J Bacteriol 183:6394–6403

Raymond KN, Dertz EA (2004) Biochemical and physical properties of siderophores. ASM Press, Washington

Rees DC, Howard JB (2000) Nitrogenase: standing at the crossroads. Curr Opin Chem Biol 4:559–566

Reigh G, O'Connell M (1993) Siderophore-mediated iron transport correlates with the presence of specific iron-regulated proteins in the outer membrane of *Rhizobium meliloti*. J Bacteriol 175:94–102

Reniere ML, Ukpabi GN, Harry SR, Stec DF, Krull R, Wright DW, Bachmann BO, Murphy ME, Skaar EP (2010) The IsdG-family of haem oxygenases degrades haem to a novel chromophore. Mol Microbiol 75:1529–1538

Rioux CR, Jordan DC, Rattray JB (1986) Iron requirement of *Rhizobium leguminosarum* and secretion of anthranilic acid during growth on an iron-deficient medium. Arch Biochem Biophys 248:175–182

Ritz D, Patel H, Doan B, Zheng M, Aslund F, Storz G, Beckwith J (2000) Thioredoxin 2 is involved in the oxidative stress response in *Escherichia coli*. J Biol Chem 275:2505–2512

Roberts GP, Youn H, Kerby RL (2004) CO-sensing mechanisms. Microbiol Mol Biol Rev 68:453–473

Robinson NJ, Procter CM, Connolly EL, Guerinot ML (1999) A ferric-chelate reductase for iron uptake from soils. Nature 397:694–697

Rodionov DA, Gelfand MS, Todd JD, Curson AR, Johnston AW (2006) Computational reconstruction of iron- and manganese-responsive transcriptional networks in alpha-Proteobacteria. PLoS Comput Biol 2:e163

Ruangkiattikul N, Bhubhanil S, Chamsing J, Niamyim P, Sukchawalit R, Mongkolsuk S (2012) *Agrobacterium tumefaciens* membrane-bound ferritin plays a role in protection against hydrogen peroxide toxicity and is negatively regulated by the iron response regulator. FEMS Microbiol Lett 329:87–92

Rudolph G, Hennecke H, Fischer HM (2006a) Beyond the Fur paradigm: iron-controlled gene expression in rhizobia. FEMS Microbiol Rev 30:631–648

Rudolph G, Semini G, Hauser F, Lindemann A, Friberg M, Hennecke H, Fischer HM (2006b) The Iron control element, acting in positive and negative control of iron-regulated *Bradyrhizobium japonicum* genes, is a target for the Irr protein. J Bacteriol 188:733–744

Salahudeen AA, Thompson JW, Ruiz JC, Ma HW, Kinch LN, Li Q, Grishin NV, Bruick RK (2009) An E3 ligase possessing an iron-responsive hemerythrin domain is a regulator of iron homeostasis. Science 326:722–726

Sangwan I, O'Brian MR (1992) Characterization of δ-aminolevulinic acid formation in soybean root nodules. Plant Physiol 98:1074–1079

Sangwan I, Small SK, O'Brian MR (2008) The *Bradyrhizobium japonicum* Irr protein is a transcriptional repressor with high-affinity DNA-binding activity. J Bacteriol 190:5172–5177

Santos R, Bocquet S, Puppo A, Touati D (1999) Characterization of an atypical superoxide dismutase from *Sinorhizobium meliloti*. J Bacteriol 181:4509–4516

Schauer K, Rodionov DA, de Reuse H (2008) New substrates for TonB-dependent transport: do we only see the 'tip of the iceberg'? Trends Biochem Sci 33:330–338

Schmitt MP (1997) Utilization of host iron sources by *Corynebacterium diphtheriae*: identification of a gene whose product is homologous to eukaryotic heme oxygenases and is required for acquisition of iron from heme and hemoglobin. J Bacteriol 179:838–845

Schuller DJ, Wilks A, Ortiz de Montellano PR, Poulos TL (1999) Crystal structure of human heme oxygenase-1. Nat Struct Biol 6:860–867

Schuller DJ, Zhu W, Stojiljkovic I, Wilks A, Poulos TL (2001) Crystal structure of heme oxygenase from the gram-negative pathogen *Neisseria meningitidis* and a comparison with mammalian heme oxygenase-1. Biochemistry 40:11552–11558

Schwartz CJ, Giel JL, Patschkowski T, Luther C, Ruzicka FJ, Beinert H, Kiley PJ (2001) IscR, an Fe-S cluster-containing transcription factor, represses expression of *Escherichia coli* genes encoding Fe-S cluster assembly proteins. Proc Natl Acad Sci U S A 98:14895–14900

Sedlacek V, van Spanning RJ, Kucera I (2009) Ferric reductase A is essential for effective iron acquisition in *Paracoccus denitrificans*. Microbiology 155:1294–1301

Silhavy TJ, Kahne D, Walker S (2010) The bacterial cell envelope. Cold Spring Harb Perspect Biol 2:a000414

Singleton C, White GF, Todd JD, Marritt SJ, Cheesman MR, Johnston AW, Le Brun NE (2010) Heme-responsive DNA binding by the global iron regulator Irr from *Rhizobium leguminosarum*. J Biol Chem 285:16023–16031

Skaar EP, Gaspar AH, Schneewind O (2004) IsdG and IsdI, heme-degrading enzymes in the cytoplasm of *Staphylococcus aureus*. J Biol Chem 279:436–443

Skaar EP, Gaspar AH, Schneewind O (2006) *Bacillus anthracis* IsdG, a heme-degrading monooxygenase. J Bacteriol 188:1071–1080

Skorupska A, Deryło M, Lorkiewicz Z (1989) Siderophore production and utilization by *Rhizobium trifolii*. Biometals 2:45–49

Small SK, O'Brian MR (2011) The *Bradyrhizobium japonicum frcB* gene encodes a diheme ferric reductase. J Bacteriol 193:4088–4094

Small SK, Puri S, O'Brian MR (2009a) Heme-dependent metalloregulation by the iron response regulator (Irr) protein in *Rhizobium* and other Alpha-proteobacteria. Biometals 22:89–97

Small SK, Puri S, Sangwan I, O'Brian MR (2009b) Positive control of ferric siderophore receptor gene expression by the Irr protein in *Bradyrhizobium japonicum*. J Bacteriol 191:1361–1368

Smith MJ, Neilands JB (1984) Rhizobactin, a siderophore from *Rhizobium meliloti*. J Plant Nutr 7:449–458

Smith MJ, Shoolery JN, Schwyn B, Holden I, Neilands JB (1985) Rhizobactin, a structurally novel siderophore from *Rhizobium meliloti*. J Am Chem Soc 107:1739–1743

Sobota JM, Imlay JA (2011) Iron enzyme ribulose-5-phosphate 3-epimerase in *Escherichia coli* is rapidly damaged by hydrogen peroxide but can be protected by manganese. Proc Natl Acad Sci U S A 108:5402–5407

Spaink HP (2000) Root nodulation and infection factors produced by rhizobial bacteria. Annu Rev Microbiol 54:257–288

Stevens JB, Carter RA, Hussain H, Carson KC, Dilworth MJ, Johnston AW (1999) The *fhu* genes of *Rhizobium leguminosarum*, specifying siderophore uptake proteins: *fhuDCB* are adjacent to a pseudogene version of *fhuA*. Microbiology 145(Pt 3):593–601

Stojiljkovic I, Baumler AJ, Hantke K (1994) Fur regulon in gram-negative bacteria. Identification and characterization of new iron-regulated *Escherichia coli* genes by a fur titration assay (published erratum appears in J Mol Biol 1994 Jul 15; 240(3):271). J Mol Biol 236:531–545

Stojiljkovic I, Hantke K (1992) Hemin uptake system of *Yersinia enterocolitica*: similarities with other TonB-dependent systems in gram-negative bacteria. EMBO J 11:4359–4367

Tang C, Robson AD, Dilworth MJ (1990) The role of iron in nodulation and nitrogen fixation in *Lupinus angustifolius* L. New Phytol 114:173–182

Tao K, Makino K, Yonei S, Nakata A, Shinagawa H (1989) Molecular cloning and nucleotide sequencing of *oxyR*, the positive regulatory gene of a regulon for an adaptive response to oxidative stress in *Escherichia coli*: homologies between OxyR protein and a family of bacterial activator proteins. Mol Gen Genet 218:371–376

Taulé C, Zabaleta M, Mareque C, Platero R, Sanjurjo L, Sicardi M, Frioni L, Battistoni F, Fabiano E (2012) New Betaproteobacterial Rhizobium strains able To efficiently nodulate *Parapiptadenia rigida* (Benth.) Brenan. Appl Environ Microbiol 78:1692–1700

Tartaglia LA, Storz G, Ames BN (1989) Identification and molecular analysis of oxyR-regulated promoters important for the bacterial adaptation to oxidative stress. J Mol Biol 210:709–719

Todd JD, Sawers G, Johnston AW (2005) Proteomic analysis reveals the wide-ranging effects of the novel, iron-responsive regulator RirA in *Rhizobium leguminosarum* bv. viciae. Mol Genet Genomics 273:197–206

Todd JD, Sawers G, Rodionov DA, Johnston AW (2006) The *Rhizobium leguminosarum* regulator IrrA affects the transcription of a wide range of genes in response to Fe availability. Mol Genet Genomics 275:564–577

Todd JD, Wexler M, Sawers G, Yeoman KH, Poole PS, Johnston AW (2002) RirA, an iron-responsive regulator in the symbiotic bacterium *Rhizobium leguminosarum*. Microbiology 148:4059–4071

Tottey S, Rich PR, Rondet SAM, Robinson NJ (2001) Two menkes-type ATPases supply copper for photosynthesis in *Synechocystis* PCC 6803. J Biol Chem 276:19999–20004

Turner SL, Young JP (2000) The glutamine synthetases of rhizobia: phylogenetics and evolutionary implications. Mol Biol Evol 17:309–319

Vashisht AA, Zumbrennen KB, Huang X, Powers DN, Durazo A, Sun D, Bhaskaran N, Persson A, Uhlen M, Sangfelt O, Spruck C, Leibold EA, Wohlschlegel JA (2009) Control of iron homeostasis by an iron-regulated ubiquitin ligase. Science 326:718–721

Verma DPS, Nadler KD (1984) The Rhizobium-legume symbiosis: the host's point of view. Springer-Verlag, New York

Viguier C, Cuiv PO, Clarke P, O'Connell M (2005) RirA is the iron response regulator of the rhizobactin 1021 biosynthesis and transport genes in *Sinorhizobium meliloti* 2011. FEMS Microbiol Lett 246:235–242

Wandersman C, Delepelaire P (2004) Bacterial iron sources: from siderophores to hemophores. Annu Rev Microbiol 58:611–647

Wandersman C, Stojiljkovic I (2000) Bacterial heme sources: the role of heme, hemoprotein receptors and hemophores. Curr Opin Microbiol 3:215–220

Wang S, Wu Y, Outten FW (2011) Fur and the novel regulator YqjI control transcription of the ferric reductase gene *yqjH* in *Escherichia coli*. J Bacteriol 193:563–574

Wexler M, Todd JD, Kolade O, Bellini D, Hemmings AM, Sawers G, Johnston AW (2003) Fur is not the global regulator of iron uptake genes in *Rhizobium leguminosarum*. Microbiology 149:1357–1365

Wexler M, Yeoman KH, Stevens JB, de Luca NG, Sawers G, Johnston AW (2001) The *Rhizobium leguminosarum tonB* gene is required for the uptake of siderophore and haem as sources of iron. Mol Microbiol 41:801–816

Wilderman PJ, Sowa NA, FitzGerald DJ, FitzGerald PC, Gottesman S, Ochsner UA, Vasil ML (2004) Identification of tandem duplicate regulatory small RNAs in *Pseudomonas aeruginosa* involved in iron homeostasis. Proc Natl Acad Sci U S A 101:9792–9797

Wilhelm SW, Trick CG (1994) Iron-limited growth of cyanobacteria: multiple siderophore production is a common response. Limnol Oceanogr 39:1979–1984

Wilks A, Schmitt MP (1998) Expression and characterization of a heme oxygenase (Hmu O) from *Corynebacterium diphtheriae*. Iron acquisition requires oxidative cleavage of the heme macrocycle. J Biol Chem 273:837–841

Williams PH (1979) Novel iron uptake system specified by ColV plasmids: an important component in the virulence of invasive strains of *Escherichia coli*. Infect Immun 26:925–932

Winkelmann G (2007) Ecology of siderophores with special reference to the fungi. Biometals 20:379–392

Wu R, Skaar EP, Zhang R, Joachimiak G, Gornicki P, Schneewind O, Joachimiak A (2005) *Staphylococcus aureus* IsdG and IsdI, heme-degrading enzymes with structural similarity to monooxygenases. J Biol Chem 280:2840–2846

Yamanaka K, Ishikawa H, Megumi Y, Tokunaga F, Kanie M, Rouault TA, Morishima I, Minato N, Ishimori K, Iwai K (2003) Identification of the ubiquitin-protein ligase that recognizes oxidized IRP2. Nat Cell Biol 5:336–340

Yang J, Ishimori K, O'Brian MR (2005) Two heme binding sites are involved in the regulated degradation of the bacterial iron response regulator (Irr) protein. J Biol Chem 280:7671–7676

Yang J, Kim KD, Lucas A, Drahos KE, Santos CS, Mury SP, Capelluto DG, Finkielstein CV (2008) A novel heme-regulatory motif mediates heme-dependent degradation of the circadian factor period 2. Mol Cell Biol 28:4697–4711

Yang J, Panek HR, O'Brian MR (2006a) Oxidative stress promotes degradation of the Irr protein to regulate haem biosynthesis in *Bradyrhizobium japonicum*. Mol Microbiol 60:209–218

Yang J, Sangwan I, Lindemann A, Hauser F, Hennecke H, Fischer HM, O'Brian MR (2006b) *Bradyrhizobium japonicum* senses iron through the status of haem to regulate iron homeostasis and metabolism. Mol Microbiol 60:427–437

Yang J, Sangwan I, O'Brian MR (2006c) The *Bradyrhizobium japonicum* Fur protein is an iron-responsive regulator in vivo. Mol Genet Genomics 276:555–564

Yeo WS, Lee JH, Lee KC, Roe JH (2006) IscR acts as an activator in response to oxidative stress for the *suf* operon encoding Fe-S assembly proteins. Mol Microbiol 61:206–218

Yeoman KH, Curson AR, Todd JD, Sawers G, Johnston AW (2004) Evidence that the *Rhizobium* regulatory protein RirA binds to cis-acting iron-responsive operators (IROs) at promoters of some Fe-regulated genes. Microbiology 150:4065–4074

Yeoman KH, Wisniewski-Dye F, Timony C, Stevens JB, deLuca NG, Downie JA, Johnston AW (2000) Analysis of the *Rhizobium leguminosarum* siderophore-uptake gene *fhuA*: differential expression in free-living bacteria and nitrogen-fixing bacteroids and distribution of an *fhuA* pseudogene in different strains. Microbiology 146(Pt 4):829–837

Yu C, Genco CA (2012) Fur-mediated activation of gene transcription in the human pathogen *Neisseria gonorrhoeae*. J Bacteriol 194:1730–1742

Zenke-Kawasaki Y, Dohi Y, Katoh Y, Ikura T, Ikura M, Asahara T, Tokunaga F, Iwai K, Igarashi K (2007) Heme induces ubiquitination and degradation of the transcription factor Bach1. Mol Cell Biol 27:6962–6971

Zheng M, Aslund F, Storz G (1998) Activation of the OxyR transcription factor by reversible disulfide bond formation. Science 279:1718–1721

Zheng M, Doan B, Schneider TD, Storz G (1999) OxyR and SoxRS regulation of *fur*. J Bacteriol 181:4639–4643

Zheng M, Wang X, Templeton LJ, Smulski DR, LaRossa RA, Storz G (2001) DNA microarray-mediated transcriptional profiling of the *Escherichia coli* response to hydrogen peroxide. J Bacteriol 183:4562–4570

Zhu W, Hunt DJ, Richardson AR, Stojiljkovic I (2000a) Use of heme compounds as iron sources by pathogenic neisseriae requires the product of the *hemO* gene. J Bacteriol 182:439–447

Zhu W, Wilks A, Stojiljkovic I (2000b) Degradation of heme in gram-negative bacteria: the product of the *hemO* gene of *Neisseriae* is a heme oxygenase. J Bacteriol 182:6783–6790